FOREWORD

The accelerating speed of depletion of tropical forests has been discussed on many occasions such as the recent UN Conference on Environmental and Development in Brazil. Since these forests are our lifeline providing oxygen, some people have suggested that they be kept intact under all circumstances. Of course, it would be quite difficult to keep them virgin because of socio-economic pressures from both producing and consuming countries. Hence we need to find effective ways and means which would give economic benefit to both parties and which are at the same time sound from the viewpoint of environment. In developed countries, trees are being used but are not so quickly depleted because of afforestation efforts and skills. Hence developed countries, which rely upon tropical forests, should share not only the cost of preserving them but also that of forestry management technology which is extremely important to alleviate this global issue. Japan and other developed countries have a lot to contribute to this field because they have learned costly lessons in the process of achieving the sustainable development. For example, Japan once suffered from environmental havocs as were typically illustrated by the Minamata or Itai-Itai diseases. Now it has accumulated technologies and experience which are relevant to reconcile development with environmental protection. The transfer of these technologies and experience from Japan has assigned an increasing significance to neighbouring countries.

These considerations led the APO to organize a Study Meeting on New Trends in Environmental Management: Forestry Resources Development and Protection, in Japan in early February 1992. The present publication is a report of the Meeting and includes the papers presented by participants and resource persons, together with the background and discussion notes. The Asian Productivity Organization hopes that this publication will be of use to all those who are concerned with environmental management in general and the development and protection of forestry resources in particular.

Kenichi Yanagi
Secretary-General
Asian Productivity Organization

Tokyo, March 1993

Table of Contents

Foreword

I. Resource Papers

PROTECTION OF JAPANESE FOREST AND TRANSFER OF THE JAPANESE EXPERIENCE

Takehiko Ohta
Department of Forestry
The University of Tokyo
Tokyo, Japan

1.   Introduction

While the forest has an economic function to supply timber and other forest products, it also has public benefit functions such as the prevention of landslides in mountainous areas, watershed conservation, conservation of the natural environment, provision of recreational sites, provision of a living environment for wildlife, and so on. Along with progress in urbanization and improvement of the living standard, the requirement for these public benefit functions of the forest is rising rapidly. Recent expansion of forest destruction and people's interest in global environmental problems accelerate such a requirement.

However, the destruction or disturbance of forests did not begin until very recently. In Japan such disturbances have increased the threat to human life by increasing the occurrence of natural disasters. Therefore, people have long been interested in the protection of forests. Usually "protection of the forest" means protecting it from climate damage, fire, pests, disease, and so on. This paper describes protecting Japanese forests from destruction or disturbance by various human activities.

2.   Natural Conditions of Japan and its Forests

Topography and geology

The Japanese islands form part of an arc-shaped archipelago that lies off the east coast of the Eurasian continent. Japanese terrain is complex and strongly influenced by a complex geological structure. The topographical and geological features of Japan are as follows:

1)   There are high mountains in the centre of the islands and deep sea trenches off the east coast of the islands.

2)   Most mountains have steep slopes and the rivers are generally short.

3   Japan has a geologically complex structure, that is, it has many type of rocks ranging in age from Paleozoic to Quaternary and it also has many faults and fractured zones.

4) Japan has many volcanoes.

5) Earthquakes frequently occur near the Japanese islands.

These geological and topographical characteristics of Japan can be explained by the plate tectonic theory. According to the plate tectonic theory, two oceanic plates (Pacific plate and Philippine plate) dive beneath the Japanese islands that occupy small parts of continental plates (Eurasian plate and North American plate), pushing them in the process. The place where the plate is subducted is the sea trench and the Japanese islands occur above the subduction zone (Fig. 1). The special arrangement of land and sea such as sea trenches, island arcs, backarc sea, and continent is a result of ocean-continent collision. The vertical displacement of the Japanese islands during the Quarternary period was very large. Such crustal movements are still in progress, and major earthquakes occur near the sea trenches. Japanese volcanoes form linear volcanic fronts about 200 km from the sea trenches.

Fig. 1 Plate tectonics near Japan

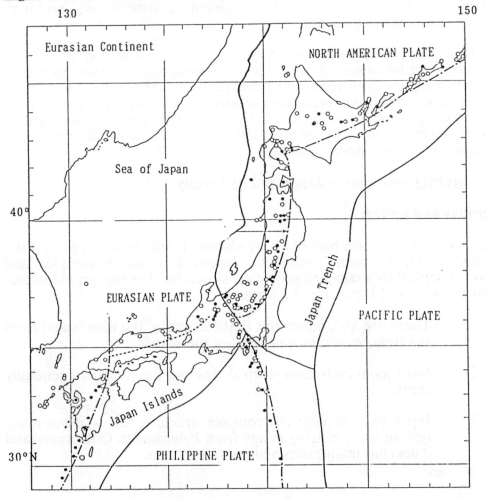

Thus, the Japanese islands belong to an extremely active mobile belt, and the geology and topography of the Japanese islands show the geological and topographical characteristics of land in tectonically active areas of the world. This means that Japanese mountains are prone to erosion. The geological reasons why they are prone to erosion can be summarized as follows:

1) Many of the mountains are composed of Tertiary and Quaternary rocks (including volcanic rocks) that were originally weak.

2) The continuous uplift in mountainous regions increases the possibility of severe erosion. (The average erosion rate in Japanese mountainous areas is approximately 10 cm/1,000 years. This erosion rate exceeds the world average by one order of magnitude).

3) Since Japanese mountains were subjected to strong compression, they were originally fractured and prone to erosion.

4) They also have many fault zones, making the mountains prone to erosion.

5) A large number of earthquakes have caused the land to become further weakened.

6) Many of the volcanoes in Japan are stratovolcanoes and contain magma of a chemical composition that causes explosive eruptions, ejecting large quantities of various volcanic products including ash and pumice into the surrounding area.

7) Large quantities of rock dust are deposited at the bases of the mountains.

Those topographical and geological features of the land simultaneously provide favourable sites and conditions for the growth of trees.

Climate and soil

The Japanese islands are located in a subtropical monsoon area and annual precipitation in Japan is 1,800 mm/year on average (Fig. 2). Precipitation occurs mainly in three periods: the rainy season, the typhoon season, and the snowfall season. Precipitation during the rainy season occurs from June to mid-July, when the Baiu front covers the southern half of Japan. At the end of the rainy season, heavy rainfall often occurs locally. In August and September three or four typhoons hit the Japanese islands on average every year, which also bring heavy rainfall. In the monsoon area in East Asia, the heavy rainfall zone in summer spreads over the middle latitudes, including most Japanese islands, being influenced by the Himalayas and Tibetan highlands. Therefore, the occurrence of typhoons in Japan and the occurrence of heavy rainfalls on the Baiu front are phenomena peculiar to such monsoon areas. Consequently, there are two or three heavy rainfalls whose total precipitation reaches more than 300 mm/day every year. These heavy rainfalls cause severe erosion in mountainous areas. On the Japan Sea side, it snows heavily in winter. Such large quantities of precipitation and the absence of a severe dry season encourages good forests.

Fig. 2. Annual rainfall (left) and annual average temperature (right).

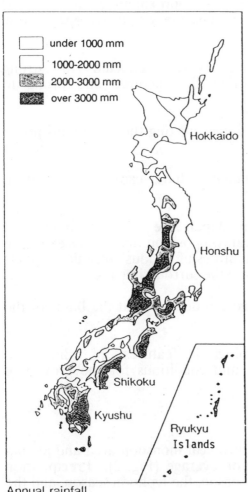

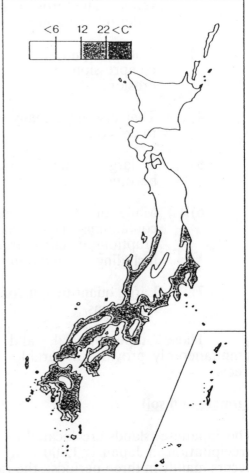

Annual rainfall   Annual average temperature

The Japanese islands are located between 26° and 46° north latitude and the annual average temperature in Tokyo (35° latitude) is 15°C. Although the northern part of Japan (Hokkaido Island) is in the subarctic zone and the southern part is in the subtropical zone, almost all the middle part belongs to the temperate zone. Therefore, most hill slopes in Japan are covered with forests, except for high alpine regions, and these slopes generally have well-developed forest soils. Forest promotes the formation of forest soil through the supply of organic materials and activity of living organisms. As Japan is located within the temperate zone, the rate of decomposition of organic materials is not as high as in the torrid zone and not as low as in the frigid zone. As a result, mature forest soils whose profiles can be divided into a few horizons are formed. This also provides favourable sites and conditions for the growth of trees.

Fig. 3. Forest distribution.

Forest Distribution and Forest Resources

In Japan the climax vegetation is forest, reflecting the warm monsoon climate with high rainfall. However, the composition of species and the distribution of forest types differ by region due to the marked climate differences in Japan's long and narrow land area, and due to the complex differences in topography, geology, and soil. These forests are horizontally classified into five types: subfrigid zone forest, mixed coniferous and deciduous broadleaved forest, cool temperate zone forest, warm temperate zone forest, and subtropical zone forest (Fig. 3).

Japan has 25.26 million ha of forests, which can be broken down into 13.67 million ha of natural forests, 10.22 million ha of manmade forests, and 1.37 million ha of other types of forest. Although this forested land accounts for about 68% of the total land area, and Japan is regarded as a country abundant in forests, the per capita forest area is only about 0.2 ha, less than half the world average.

Fig. 4. Forest area in Japan.

Unit: 10,000 ha

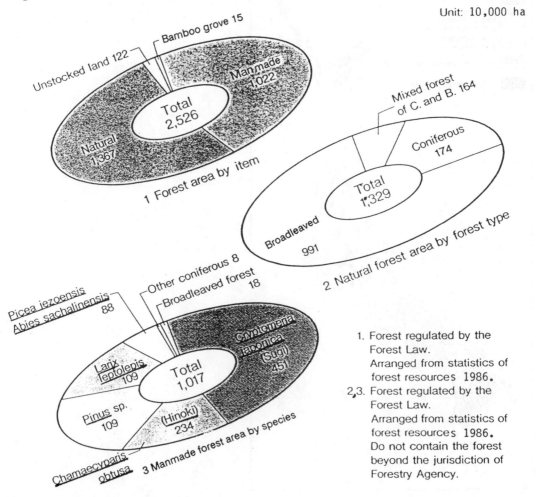

1. Forest regulated by the Forest Law.
   Arranged from statistics of forest resources 1986.
2,3. Forest regulated by the Forest Law.
   Arranged from statistics of forest resources 1986.
   Do not contain the forest beyond the jurisdiction of Forestry Agency.

The growing stock of natural forests is about 1.5 billion m^3 while that of manmade forests is about 1.36 billion m^3 (Fig. 4). The volume of growth of manmade forest at present is about 76 million m^3 a year, which steadily increases the growing stock. However, the distribution of stand age classes is uneven and the majority of stands are under 35 years old (Fig. 5). Accordingly, the currently available volume is considerably less than the theoretical figure.

Fig. 5. <u>Growing stock (left) and manmade forest (right) by age class</u>.

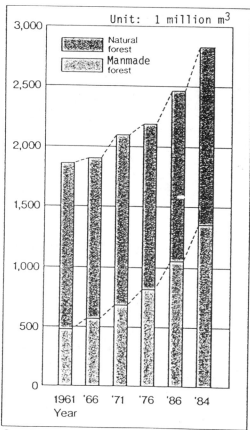

Growing stock (1986)

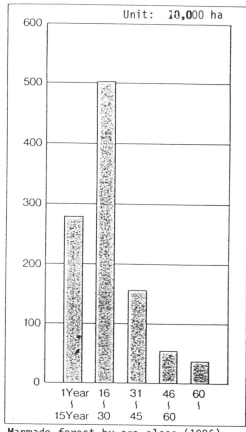

Manmade forest by age class (1986)

Fig. 6. Forest resources by ownership (1986).

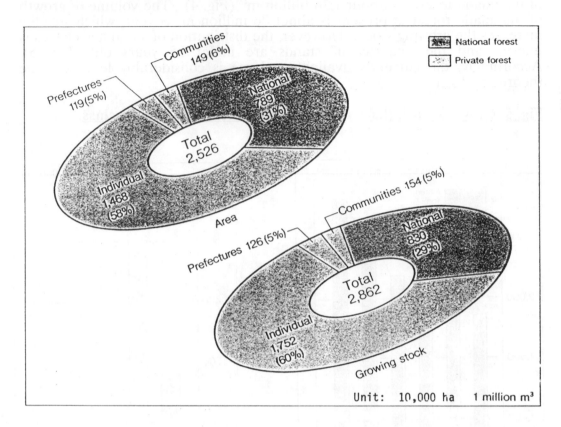

The quantity of logs currently harvested is about 13 million m^3 from natural forests and 18 million m^3 from manmade forests. By the beginning of the 21st century, when stands planted after World War II will mature, the potential of the supply is expected to increase greatly.

Fortunately, the forested land area in Japan has not decreased for the last three decades for the following reasons:

1) Reafforestation is comparatively easy because of the favourable natural conditions.

2) Most forests are located on steep mountainous slopes that cannot be developed.

3) In hilly areas near populated areas, the reafforestation efforts have been carried out for a long time.

4) A large quantity of timber is imported.

3.   Experience of Reafforestation in Japan and the Protected Forest System

History of forest disturbances in Japan

Destruction and disturbance of forest began and expanded with the advance of civilization.   Once there were widespread bare land tracts in the central part of Japan.   In ancient times, timber for court buildings in the Heijo capital (Nara) and the Heian capital (Kyoto) was obtained from neighbouring areas.   As a result, the mountains near Kyoto and Nara were eroded due to the indiscriminate felling of trees.   Since then felling has continued in suburban forests to obtain building and fuel materials or fertilizer for rice fields.   Consequently, disasters such as debris slides and debris flows occurred in mountainous areas and severe floods occurred frequently on downstream plains.

The government often prohibited or restricted the felling of trees in order to prevent erosion.   For example, in 1682 the feudal government prohibited cutting of headwater forest and encouraged planting in the Yodogawa River basin. (In 1689, check dam construction was initiated.) But the situation continued until the end of the 19th century.

In 1896 and 1897 the government promulgated the River Law, the Forest Law, and the Sabo Law (Sediment Control Law), in order to carry out erosion and sediment control projects throughout the country.   These three laws are called the Three Laws for Flood and Sediment Control.   From then on, forests in suburban areas began to improve.   Today's situation of forested land in Japan is better than in the past.

Function of forest for erosion control

When discussing erosion control in mountainous areas, we must consider the function of the forest for erosion prevention.   The reasons why forests prevent erosion are:

1)   The soil surface is covered with organic materials (A_0 horizon or forest floor vegetation), which protect the soil surface from erosion.

2)   Forest soil contains tree root system, which often reach the bedrock and make the soil stronger.

3)   The forest soil is very permeable and is a structural soil with varying porosity.   As a result, rainwater stored in the soil is drained rapidly during and after rainfall. (Pore pressure in the soil does not increase.)   This means that it is more difficult for debris slides to occur.

As described above, forest cover prevents surface erosion on hill slopes.   Instead, mass slides predominate on forested hill slopes.   Mass slides are classified into two types: debris slides and bedrock slides. Debris slides are characterized by sliding of weathered materials that mainly consist of forest soil on the bedrock. Bedrock slide materials consist of fractured bedrock and surface soil. The forest prevents debris slides.

The Japanese have traditionally realized these functions of the forest based on their experience from historical times and have established the "protected forest system" as a method of erosion prevention.

Protected forest system

In addition to the function of sediment disaster prevention, the forest provides additional public benefits. Since river gradients are steep and rainfall varies greatly throughout the year in Japan, the conditions for the effective use of water resources are not ideal. On the other hand, due to expanded industrial production and improved living standards, the demand for water has grown continuously. Therefore, the function of the forest to ensure a stable water supply is very important. In addition, people have been showing an increasing interest in the natural environment, and forests have become very popular places for recreation. Forests are also used for the conservation of birds and animals.

In order to preserve forest expected to fulfil such public benefit functions or to establish forests in areas that need such functions, the protected forest system has been established based on the Forest Law promulgated in 1897. Forests are registered as protected forests in accordance with the law and are placed under certain limitations in management.

Three restrictions are mainly imposed on the management of protected forest: restriction on felling of standing trees, conversion of topography, and obligation to plant. Felling of standing trees is to be done within the framework of the Forest Management Plan that provides for a felling system (clearcutting, selective cutting, or cutting prohibition) and the annual cutting area, and the conversion of land topography is regulated so as to require the permission of the governor concerned. After felling, the regulation requires planting to commence within two years.

If owners of protected forests are individuals, since private property is restricted for use for the public service, the government grants the following special favours: tax exemption and reduction; financing; raising of reafforestation subsidies; in case of natural disasters, execution of planting at the expense of the government (without any burden on the individual); availability of profit-sharing reafforestation with public entity; and monetary compensation for loss. Forest resources by ownership are shown in Fig. 6.

The protected forest system has been modified several times since 1897. In 1989, the protected forests were divided into 16 types according to purpose (Table 1). The protected forest system has been effective in preserving forests. It is fortunate that this excellent system was established some 100 years ago.

Table 1.   **Protected forest area by type (1989).**

Type	Area (ha)
Water conservation	5,966,264
Erosion control	1,868,260
Soil conservation	45,141
Shifting sand control	16,357
Windbreak	54,714
Flood damage prevention	742
Tidewater damage prevention	13,143
Drought damage prevention	38,195
Snowbreak	-
Fog prevention	51,323
Avalanche prevention	19,103
Falling rock prevention	1,690
Fire prevention	407
For fish	27,827
Target for navigation	1,105
For health (recreational use)	544,057
For scenic beauty	27,796

4. Erosion Control Works in Japan

Social conditions in Japan and sediment-related disasters

The population of Japan is about 120 million. About 80% of the population is concentrated on the plains, which account for only 24% of the total land area (370,000 km^2) (Fig. 7). Therefore, the population density in habitable areas is very high and intensive land use practices are common. For this reason, even dangerous areas are crowded with houses near cities. On the other hand, in rural areas land use changes, such as road construction and development of golf courses and ski resorts, are in progress. Among them, the most drastic land use change is the shift from forested land to residential or industrial areas.

Fig. 7. Distribution of Japanese population by land type.

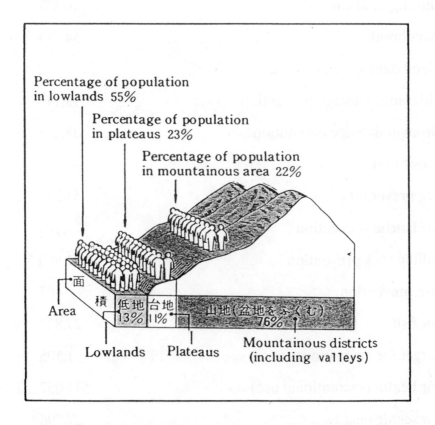

Due to the natural conditions explained in section 2 and the social conditions described above, various kinds of sediment-related disasters frequently occur in Japan. For example, on steep slopes two types of slides occur, on average, once or twice every few years. Sliding mainly occurs during the rainy and typhoon seasons. Slides that occur during heavy rainfall are usually debris slides, while bedrock slides are quite rare. Therefore, debris slides are the most predominant type of erosion in Japan, in particular, in the granitic and Tertiary areas.

Both types of slide frequently become mudflow (or debris flow) torrents, which cause serious disasters on downstream fans. In Japan, most small fans at the bases of hill slopes are composed of mudflow and gravel flow deposits. Therefore, debris slides cause serious problems and the function of forests with good tree root systems to prevent them is very important. The only protected forest system alone cannot provide complete prevention. There are also occasional mudflows or lava flows (or even pyroclastic flows) in volcanic areas. Moreover, a large-scale collapse occurs about once every 10 years. To prevent themselves from such disasters, the Japanese have taken great interest in sediment disaster preventive works going back to historical times, and have carried out many kinds of erosion or sediment control works. Such erosion control engineering is referred as "sabo" engineering.

Erosion control works

In Japan, projects for prevention of sediment-related disasters are undertaken by the Ministry of Construction and the Forestry Agency, a lower organization of the Ministry of Agriculture, Forestry and Fisheries. The works undertaken by the Ministry of Construction are called sabo works (sediment disaster preventive works), and those undertaken by the Forestry Agency, forest conservation works. The former have as their purpose the protection of human living and property in basins, while the latter have as their major aim the conservation of forests.

Sabo works

The execution of sabo works is mainly carried out in accordance with the Sabo Law. Sabo works include sediment control projects, construction to prevent landslides, and additional construction to prevent steep slope failures. Their purposes can be further classified as follows:

1)　　sabo works for drainage systems, meaning sediment control of the overall basin;

2)　　sabo works for local areas, including disaster preventive measures in small piedmont areas, preventive measures against debris flows, and sabo works in urban areas;

3)　　preventive measures for landslides;

4) preventive measures related to sedimentation in reservoirs;

5) improvement of raised-bed rivers;

6) environmental improvement projects; and

7) <u>sabo</u> works for active volcanoes.

Forest conservation works

Forest conservation works are mainly executed in accordance with the Forest Law. The purposes of forest conservation measures can be classified into the following categories: 1) soil conservation works, including soil conservation works for reafforestation and prevention, intensive conservation works for mountain disaster hazard areas, intensive conservation works for river-head areas, and landslide prevention works; and 2) protected forest maintenance works consisting of (ordinary) protected forest improvement works, protected forest improvement works for a desirable living environment, and establishment of disaster prevention forests.

Generally speaking, forests in Japan today maintained in good condition. This is due to the excellent protected forest system based on the Forest Law. The land area of protected forests reaches about 8 million ha.

5. Forest Administration Concerning Protection of Forests

Forestry Agency

The central organization of forest administration is the Forestry Agency, a lower organization of the Ministry of Agriculture, Forestry and Fisheries. With the aim of ensuring the function of forests of the public benefit and a stable wood supply by developing healthy forests, the Forestry Agency is in charge of managing national forests and guiding and assisting private and public forestry and forest industries through local government (Fig. 8). This is achieved by planning forestry policies, making budgets and settling accounts, supporting forest conservation programmes, construction of forest roads, reafforestation, thinning, and structural improvement programmes.

Forest policy

In Japan, among the laws related to the forests and forestry, the Forest Law, the Forestry Basic Law, and the Forestry Cooperative Association Law are the most significant. (The first law included the third law before 1978). Since the first Forest Law was established in 1897, this law has a long history, and was drastically revised in 1907, 1939, and 1951. After that is was also often revised. The current Forest Law was promulgated in 1983. The aim of this law is to achieve continuous maintenance of forests and to increase their production potential, so as to contribute to conservation of the national land and development of the national economy. This law provides for the supervision of management of forests, the maintenance and improvement of protected forests and protected facilities, and the establishment of forest plans and forest management plans.

The Minister of Agriculture, Forestry and Fisheries, in order to conserve the national land through prevention of debris slides and protection of water storage areas or conservation of sites of scenic beauty and historical interest, is authorized to register forests as protected forests or protected facilities areas. Within such registered areas, any project needed for formation or maintenance of such forests is allowed. A person who is going to fell standing trees or collect stones from such registered protected forests or protected facilities is required to obtain permission from the local governor, and the forest owner has the responsibility to plant trees after cutting.

The Forestry Basic Law was established to develop forestry in Japan and to raise the social status of all parties engaging in forestry, as well as securing forest resources and conserving the national land. As basic measures to accomplish these aims, the law stipulates:

1)   enhancement of forestry use of mountainous areas;

2)   modernization of forestry management by grouping forest fields, mechanization of forestry operations, and enlargement of management scale;

3)   improvement of forestry technologies;

4)   stabilization of the supply of and demand for forest products and their prices, and rationalization of the distribution and processing of the products;

5)   promotion and recruitment of personnel qualified to engage in modern forestry management and technology; and

6)   improvement of the welfare of forest workers.

Along with rising requirements of national life, requirements for the public benefits of forests are always rising. Therefore, the Forestry Basic Law also provides that full consideration should be paid to securing the national land and other functions of public benefits which the forest possesses. When forests needed for conservation of headwater and national land safeguards are registered as protected forests, such forests, in coordination with timber production, are planed under certain limitations in management. Thus the function of the public benefits of the forests is assured, and erosion control works are executed for conservation of the national land simultaneously.

Fig.8. Organization of the Forestry Agency.

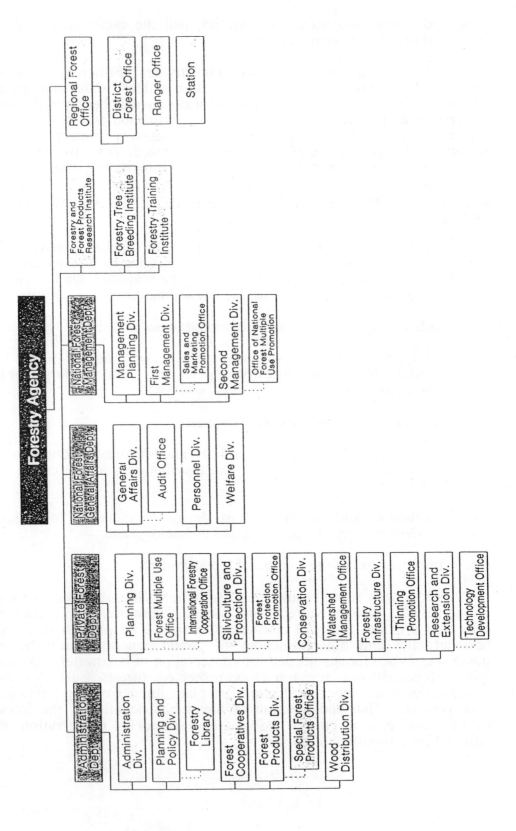

Since the forest has an important function of national land safeguard to conserve water reservoirs and prevent floods, the Erosion and Flood Control Emergency Measures Law was formulated in 1960 and the Protected Forest Fortification Temporary Measures Law in 1954. The former provides that in order to coordinate soil and water conservation projects and promote their urgent and systematic execution, the Ministers of Agriculture, Forestry and Fisheries, and Construction formulate and implement the Soil Conservation Five-Year Plan and the Water Conservation Five Year Plan starting from 1968, respectively. The latter provides that for the purpose of urgent fortification of protected forest, the government must formulate the Protected Forest Fortification Plan. Based on this plan, the nationwide Forest Plan will be changed and the government will purchase forest that is then designated as protected forest.

In addition, the Forest Land Development Sanction System was established for all forests in 1974, in order to protect forested land from being rushed into development. Under this system, approval from the relevant governors is required when developing forest land.

In recent years, forests have been used by more people in many different ways. The national and prefectural governments and therefore promoting the designation of forests as natural parks and recreational forests so that as many people as possible can use them. Efforts are also being made to conserve scenic views and improve facilities like hiking trails and camping areas. There are national parks, quasinational parks, and prefectural national parks, and their total space is 5,196,000 ha (13% of the total land space) as of 1978. Twenty-two national forests (113,000 ha) are designated as natural recreational forests and are open to the public. In addition, there are areas designated as animal protection districts totalling 2,828,000 ha to protect wildlife.

Besides the laws related to forest administration directly carried out by the Forestry Agency, laws that include measures related to forest protection are as follows:

1) National Park Law (1957);

2) Natural Environment Protection Law (1972);

3) Wildlife Protection Law (1922);

4) Urban Green Space Conservation Law (1973);

5) National Land Use Plan Law (1974);

6) Productive Green Space (Conservation) Law (1974);

7) Cultural Properties Protection Law (1950);

8)   Law to Protect Historic Sites, Places of Scenic Beauty, and Natural Monuments; and

9)   Ancient City Preservation Law.

6.   <u>Transfer of Japanese Experience</u>

<u>Overseas technical cooperation in the field of forestry</u>*

The forests of the world are decreasing and being devastated rapidly due to excessive shifting cultivation, expansion of grazing, increased harvesting of fuel wood, and damage from acid rain. Especially noticeable is the decrease in tropical forests which account for about half of the world's forest area. The decrease is estimated to be 11.3 million ha per year and according to recent reports it has accelerated to 17 million ha per year. There is fear that this decrease in tropical forests will have deleterious effects on the global environment and ultimately threaten the very existence of humans, not to mention the serious effects on local inhabitants. Economic and technical cooperation for the conservation and afforestation of tropical forests has thus been addressed in various international conferences.

Under these circumstances, Japan has been cooperating with many developing countries in the conservation and utilization of forests, afforestation, and other projects, by fully utilizing the knowledge and techniques gained through the long history of forestry in Japan which includes experience in reafforestation of manmade forests of over 10 million ha.

The method of cooperation is called "project type cooperation" in which dispatch of experts to the country requesting the cooperation, training of trainees in Japan, and provision of machines and equipment such as tractors are combined and implemented for each project. Wide-ranging cooperation activities include the development and transfer of afforestation techniques, technical guidance and the training of foresters in relation to forest ecology, and research on high-level wood utilization. In addition, Japan receives about 100 forestry trainees every year. Eight technical cooperation projects in seven countries have been completed so far. Fifteen projects in 12 countries in Eastern Asia, South America, and Africa are currently underway and a total of about 70 long-term experts despatched are now working in these countries.

There are also other types of cooperation, which include 1) cooperation in various surveys such as forest inventory; 2) grant aid for providing equipment and facilities for the development and improvement of technology and research; and 3) provision of funds.

------

* This section is taken from "Forestry and forest industries in Japan" edited by the Japan Forest Technical Association in 1991. Therefore, this includes technical cooperation not only related to forest protection but also in other fields of forestry.

Other than the above, Japan is financing activities such as the conservation, utilization, and reafforestation of tropical forests conducted by international agencies including the International Tropical Timber Organization (ITTO) and FAO. ITTO was established to deal with various problems related to tropical forests and wood through cooperation between producing and consuming countries. With its headquarters in Yokohama, Japan, ITTO receives its largest single contribution from this country. Japan is also financing FAO activities such as the Tropical Forest Action Plan which is implemented by various countries in the conservation and afforestation of tropical forests, and is taking part directly in planning with seven countries.

Technical cooperation with foreign countries in sabo works

Japan has a great deal of experience in soil and water conservation and erosion control works. Besides the Forestry Agency, the Ministry of Construction has also been cooperating with many countries in various kinds of sabo works for a long time. Especially, in volcanic sabo works, active technical cooperation in pan-Pacific Rim countries is taking place. For example, Indonesia, the most prominent volcanic country, has experienced many volcanic disasters. A Volcanic Sabo Technical Centre was established as the centre of study of active volcano sabo works where Indonesian and Japanese engineers are working together for technical development.

Bibliography

1)   Forestry Agency. Forestry in Japan.

2)   Forestry Agency. Forestry Conservation in Japan. 1987.

3)   Forestry Agency. Forestry and Forest Industries in Japan. 1991.

4)   Japan Society of Erosion Control Engineering. Erosion Control Works in Japan. 1985.

5)   Eco-systems. 1991.

GLOBAL ENVIRONMENTAL PROBLEMS AND ASIAN DEVELOPMENT PATTERNS

Ryuichiro Matsubara
College of Arts & Sciences
University of Tokyo
Tokyo, Japan

1. Economics and Global Environmental Problems

The science of economics has formulated many theoretical tools in an attempt to discover economic systems more friendly to the environment. Environmental economics on a global scale regards its first theme as the "efficiency problem." What kind of inefficiencies taken place when air, soil, or other natural resources are used? How can the most efficient allocation of these resources be attained? The second theme is the "distributive problem." What kind of distribution problems arise with the inefficient allocation of resources? What types of distribution problems might occur when one particular inefficiency is removed? And what kind of policy should be enacted to attain distributive equity?

To solve these problems, many economists have insisted that indirect regulation through the price mechanism is better than direct regulation. This point of view was established by A.C. Pigou, who considered environmental problems to be "technological externalities" and recommended tax policies to obtain socially optimal outputs (Fig 1). Pigou also pointed out that a subsidy policy could serve the same function as a tax policy (achieve socially optimal leaves of output), if distribution could be ignored. Recently, Professor J.H. Dales of Toronto University proposed the idea of a "discharge claim exchange system" (Fig. 2). He proposed making discharge claims on polluted materials exchangeable on the market. A government could issue claims for target levels of pollution, and firms that want to discharge polluted materials must buy claims to the necessary amount according to the discharge volume.

Japan experienced severe environmental pollution during the late 1960s and early 1970s. Subsequently, policies were enacted, but the majority of these countermeasures were direct regulation, or inherently "react and cure" policies. Unfortunately, many people were victimized by environmental pollution. Judging from the Japanese experience, environmental policy must be anticipated and prevented. Thus indirect rather than direct regulation is considered a more effective tool.

Recently economists have argued that both direct and indirect (tax and/or subsidy) economic policies have defects, since global environmental problems contain variables not considered by the Pigouvian externality approach. First, the effects of certain economic activities on the environment, especially the global environment, are uncertain. Consider the global warming phenomena: effects on weather patterns resulting from the increased density of CO_2 or other gases in the air caused by the greenhouse effect are still uncertain. Under direct regulation, it is impossible to estimate the social cost of CO_2 discharge to the air and hence decide the optimal discharge volume.

Second, the effects on the global environment are irreversible. Once CO_2 gas has been accumulated and weather patterns have been altered it will be impossible to return to the original, natural patterns, potentially threatening the existence of mankind. Given that the consequence of CO_2 discharge is global warming, the irreversible damage will take place after the present generation has died. Therefore, in determining optimal levels of discharge, future generations must be considered. It is our social and environmental responsibility to consider their interests.

In classical environmental problems, the causer and the sufferer are known and their nationalities are the same. Under those conditions the Pigouvian externality approach was valid. But since current environmental problems are global, the causer and the sufferer are not known and their generations and nationalities are different, especially in the case of global warming. Therefore economists must make efforts to predict transitory economic conditions that may be influential on the global environment before setting economic policies. For all the reasons mentioned above, this paper discusses the relation between the development patterns of Asian countries and global environment problems.

2.   Framework

This paper focuses on the global warming problem caused in part by CO_2 discharge. Regarding ozone depletion by chlorofluorocarbons (CFCs) consensus among the developed countries has been reached concerning the regulation of CFC use. In the case of global warming, however, the source and volume of CO_2 are difficult to monitor because they are the result of various factors, the main two of which are the consumption of fossil fuels and deforestation.

These two factors are influenced by economic conditions, particularly a country's stage of development. The present focus on Asian countries in discussion of economic development is because the Asian-Pacific region has been developing at the highest rate in the world, and has formed a closely integrated economic network. The world economy today is truly a "borderless economy" made up of closely integrated networks of private enterprises. Like the EC and American Canadian area, the Asian-Pacific region exemplifies the global economy today.

3.   Environmental Problems and Asian Development Patterns

Development patterns in the Asian-Pacific region in the late 1980s

Developing countries in East and South Asia have achieved steady growth at high rates since the 1970s. But in the early 1980s, accumulating debts became a serious problem for them and stalled their economic growth. Nonetheless, the newly industrializing economies (NIEs) made a comeback in 1986-1987 and the economies of the ASEAN nations recovered as well. After the Plaza Accord appreciation of the yen brought international reorganization to industries in the NIEs and ASEAN nations. The increase in direct investment in the ASEAN nations by Japan and NIEs established a pattern of high growth in this area, led primarily by export-oriented industries (Tables 1 and 2).

There were drastic changes in the production and export structure during this growth period. Generally speaking, production and export structures shift from agriculture to labour-intensive light industries, then to capital- and technology-intensive heavy industries, and finally to service industries. In particular, appreciation of the yen triggered direct investment, built a closer interdependent network of commerce through trade, and established a stratified form of division of labour. This pattern of development is called the "wild geese flight pattern."

Consumption of fossil fuels by development pattern

This section examines the ratio of heavy industry to the volume of air pollution gases like CO_2, ultimately expressing the relation as an index. The index that measures the effectiveness of macro energy utilization is the energy/GDP (E/G) ratio. In developed countries, the E/G ratio has decreased about 1% per year for several decades. This is because the percentage of manufacturing industry to industry as a whole reaches the maximum at some stage, and after that the ratio of service industry increases and the unit consumption of energy for unit GDP decreases due to technological progress. This energy consumption pattern has been evident in Japan since the high economic growth period from 1955. Along with the progress of heavy industry and economic growth came a rising E/G ratio and consequent burden on the environment. At the peak of the E/G ratio, environmental pollutants spread across Japan, causing Minamata disease, itai-itai disease, Yokkaichi asthma, etc. In and after the 1960s, regulations on environmental pollution were established and the E/G ratio began to decrease. The result was new economic growth without the anticipated stagnation. This was avoided because environmental regulations (prohibiting excessive gas exhaust) encouraged competition for new technologies.

In developing countries, however, the E/G ratio continued to increase. This was due to the fact that the primary energy source in agriculture changed from agricultural wastes and firewood to fossil fuels such as petroleum. But more importantly, the ratio of heavy industry to industry as a whole increased. Two measures are considered effective in curving macro energy inefficiencies in developing countries. The first is aggressive introduction of energy-conserving technologies. Although the capital investment may seem too high in the short run, it will improve economic efficiency and competitiveness in the long run, as evident in the case of Japan's economic growth. Second, it is important to reassess the conventional pattern of development from heavy industry at the intermediate stage to the postindustrial society, which appears especially applicable to the Asia-Pacific region.

Heretofore, economic development has followed the Petty-Clark theory. According to this theory, the industrial structure develops first in the primary industries (agriculture, fishery), then in the secondary industries (manufacture, construction, mining, etc.), and finally in tertiary industries (commerce, finance, services, etc). The secondary industries develop first in light and then in heavy industry. But in Asia, while following the theoretical pattern of progress, interesting characteristics in terms of energy efficiency are found in the relatively diversified patterns of development. Figure 3 shows the GDP growth in response to changes in industrial structure in Asian-Pacific countries. The graph shows that the ratio of heavy industries rises in the initial phase of economic development but after achieving the maximum point, the curve goes down as the Petty-Clark theory predicts. In studying individual countries, however, as in the case of the Republic of Korea, the placement of emphasis on heavy industries did not allow a maximum to be reached before the ratio began to decrease.

While the economy as a whole may be growing, there may be an overlapping pattern without significant difference in development among countries in the region, as in the formation of flying wild geese. Latecomers learn from the experience of early ones and can minimize the cost of technological research by following them, or being directly invested in by developed countries. This pattern was seen in NIEs but recent economic development in the region including NIEs has had other characteristics. Countries with different GNPs and different heavy industries form a network of trade and direct investments as if they share international industrial roles on the same dimension, not as in the relation between early and late entrants. The case of Korea, which has entered a new economic development stage without achieving a high ratio of heavy industry is considered to be based on such characteristics of the regional economy.

The foregoing statements become even clearer in Fig. 4, plotting the correlation between the volume of energy consumption and the rate of heavy and chemical industrialization. The basic tendency is that each country increases energy consumption in proportion to increases and/or improvements in heavy industrialization. At a certain point, however, with the increasing ratio of information and service industries to industry as a whole, individual countries begin reducing consumption as the rate of heavy industrialization diminishes. This tendency is remarkably obvious in Singapore.

Indonesia, as well as India, is in the early stage of development which promotes heavy industrialization. In energy consumption, Indonesia has levelled off while India has been rapidly expanding. One can thus understand why India's industrialization has not significantly contributed to income increase, but rather served as a source of environmental burden. India has been undergoing inefficient heavy industrialization.

Japan, as well as other advanced countries, has been reducing its rate of heavy industrialization commensurate with its energy consumption. Japan in particular has a lower rate of consumption when weighed against the level of heavy industrialization. This indicates that Japan's energy conservation technology, developed due to its experience with oil crises and environmental pollution, is at an efficient, effective level. Korea has been also changed remarkably from heavy and chemical industrialization. Korea has initiated this change at a less developed stage of heavy industrialization than has been done in the past by other developing countries, such as Singapore.

From the foregoing, it is clear that it is possible to promote economic development while minimizing the environmental pollution burden. Thus there exists a way of converting the industrial structure of a country from its early stages of heavy industrialization to a diversified structure focused on light and high-technology industries. It is possible therefore for all countries to break away from heavy industrialization. Even if countries bestowed with relatively superior energy conservation technology undertake such actions, countries that have avoided heavy industrialization can secure an opportunity for economic development in a network of direct investment and trade under the transportation and telecommunication network that the Asia-Pacific region is presently constructing.

"A new path of sustainable development" means pursuing a developmental route (recently undertaken by the NIEs, Thailand, Malaysia, etc.) on which countries can pursue their economic development through the introduction of various unconventional industries, such as light industries, tourism, etc. without concentrating on the establishment of heavy industries in their initial stages, industries that discharge large quantities of environmental pollutants such as SOX, NOX, CO_2, etc.

Decrease in forest area resulting from the Asian development pattern

Based on the wild geese flight pattern analogy, it might be assumed that countries with the least development will have a commercial forestry industry since forestry is considered a primary industry in the international sector, until industrial development and the appreciation of its currency begin. According to this theory, other countries will be able to conserve and/or recover the forest, because they have other, more sophisticated and lucrative sources of national income.

In Table 3, Thailand's recent growth rate is very high. From January 1989 the government of Thailand prohibited all types of logging. In Thailand, the percentage of forest to land was 72% in 1938, but only 28% in

1987. Not only for the environmental security of Thailand but also for the global environment this policy is reasonable and necessary. Developed countries should make efforts to conserve and recover the forest. If the entire stock of forest continues to decrease in spite of such efforts, economic intervention should be planned; for example, taxation on logging. We must ensure that the world forest is able to sustain mankind. It must be sustainable like economic development.

Table 1. <u>Foreign direct investment from Japan to other Asian countries (number, $ million, %).</u>

Country and area	1986		1987		1988		1989		1990 1st half	
	Number	Total Amount	Number	Total Amount	Number	Total Amount	Number	Total Amount	Number	Total Amount
Korea	111	436 (225.4)	166	647 (48.4)	153	483 (-25.3)	81	606 (25.5)	31	147 (-61.2)
Taiwan	178	291 (155.3)	268	367 (26.1)	234	372 (1.4)	165	494 (32.8)	48	230 (14.4)
Hong Kong	163	502 (283.2)	261	1,072 (113.5)	335	1,662 (55.0)	335	1,898 (14.2)	131	1,085 (15.4)
Singapore	85	302 (-10.9)	182	494 (63.6)	197	747 (51.2)	181	1,902 (154.6)	82	371 (-54.2)
Asian NIEs	537	1,531 (113.2)	877	2,580 (68.5)	919	3,264 (26.5)	762	4,900 (50.1)	292	1,833 (-21.3)
Indonesia	46	250 (-38.7)	67	545 (118.0)	84	586 (7.5)	140	631 (7.7)	76	442 (30.8)
Malaysia	70	158 (100.0)	64	163 (3.2)	108	387 (137.4)	159	673 (73.9)	82	373 (30.4)
Philippines	9	21 (-65.6)	18	72 (242.9)	54	134 (86.1)	87	202 (50.7)	28	93 (-7.9)
Thailand	58	124 (158.3)	192	250 (101.6)	382	859 (243.6)	403	1,276 (48.5)	204	580 (0.2)
ASEAN (4)	183	553 (7.2)	341	1,030 (86.3)	628	1,966 (90.9)	789	2,782 (41.5)	390	1,488 (14.1)
Asian total	720	2,084 (58.6)	1218	3,610 (73.2)	1547	5,230 (44.9)	1151	7,682 (46.9)	682	3,321 (-8.6)

Sources: Statistical data on report basis to the Ministry of Finance. Figures in parentheses are rates of increase in comparison to the previous year.

Table 2. Inflow of foreign direct investment to ASEAN countries.

ASEAN investor		Korea	Taiwan	Hong Kong	Singapore	USA	Japan	Total
Thailand (million baht)	1987 share	2,303 1.4	14,642 9.0	7,107 4.3	4,874 3.0	19,257 11.8	46,987 28.7	163,470 100.0
	1988 share	3,679 0.9	54,287 13.8	20,108 5.1	16,054 4.1	92,767 23.5	148,221 37.6	394,212 100.0
	1989 share	9,482 2.8	30,273 8.9	36,172 10.6	18,483 5.4	31,497 9.2	135,769 39.8	341,496 100.0
Malaysia (million ringgit)	1987 share	2 0.3	119 15.8	28 3.7	135 18.1	61 8.2	230 30.7	750 100.0
	1988 share	23 1.2	384 19.1	130 6.4	172 8.6	253 12.6	561 27.9	2,011 100.0
	1989 share	79 2.3	996 29.5	113 3.3	267 7.9	127 3.8	1,061 31.5	3,373 100.0
Philippines (million pesos)	1987 share	N.A.	186 5.4	570 16.6	N.A.	740 21.6	591 17.3	591 17.3
	1988 share	N.A.	2,306 24.2	564 5.9	N.A.	3,216 33.8	1,996 21.0	9,523 100.0

(Table 2 continued)

Indonesia 1989 share	N.A.	2,950 16.9	1,960 11.2	N.A.	2,420 13.8	3,430 19.6	17,480 100.0
1987 share	23 1.6	12 0.8	134 9.1	1 0.1	91 6.2	524 35.7	1,467 100.0
(million dollars) 1988 share	206 4.7	912 20.7	258 5.9	250 5.7	671 15.2	255 5.8	4,409 100.0
1989 share	270 9.1	59 2.0	233 7.8	55 1.8	61 2.0	484 16.2	2,981 100.0

Sources: Figures are based on each country's statistics on foreign direct investment.

Table 3. **Comparison of the real rate of economic growth by region (%).**

(1) **Asia-Pacific region**

Country	1980-1984	1985	1986	1987	1988	1989
Japan	3.9	4.9	2.5	4.6	5.7	4.9
Korea	6.6	6.9	12.4	12.0	11.8	6.1
Taiwan	7.2	4.9	11.6	12.3	7.3	7.7
Hong Kong	7.9	-0.1	11.9	13.9	7.2	2.5
Singapore	8.5	-1.6	1.8	9.4	11.1	9.2
Thailand	5.9	3.5	4.5	8.4	11.0	11.8
Malaysia	6.9	-1.0	1.2	5.3	8.7	7.6
Philippines	1.4	-4.3	1.4	4.7	6.2	6.0
Indonesia	5.0	8.7	5.9	4.8	5.7	N.A.
China	8.6	13.5	7.7	10.2	11.1	3.7
Australia	2.7	5.3	1.6	4.1	3.8	4.9
New Zealand	2.6	6.6	-3.1	1.3	1.7	0.8

(2) **North America**

Country	1980-1984	1985	1986	1987	1988	1989
USA	1.9	3.4	2.7	3.4	4.5	2.5
Canada	2.2	4.7	3.3	4.0	4.4	3.0

(3) **Europe**

Region	1980-1984	1985	1986	1987	1988	1989
E C	1.3	2.4	2.7	2.7	3.9	3.5

Sources: OECD Economic Outlook, December 1990;
ADB Key Indicators of Developing Asian and Pacific Countries, July 1990.

Bibliography

Foundation for Advanced Information and Research. Environmental problems and proposals in the Asia-Pacific region. Presented at the Second Asia-Pacific Conference on the Asia-Pacific Region in the 1990s: Cooperation for Sustainable Development and New World Order, vol. II. Tokyo: Foundation for Advanced Information and Research, 1991.

Ueda, K., Ochiai, H., Kitabatake, Y. and Teranishi, S. Kankyo Keizaigaku (Environmental economics). Tokyo: Yuhikaku Books; 1991.

Fig. 1. Private marginal cost (PMC) is the marginal cost that suppliers belonging to the private sector must pay actually. OM is external diseconomy or that cost incurred for the production of X. Social marginal cost (SMC) is MC for the society. PMC is paid and OM not actually paid.

Socially the optimal level of production is X^* but equilibrium in the market is $\hat{X}$, because PMC is the supply curve actually paid and D is the demand curve. $\hat{X}$ is more optimal X^*. If the government has tax level EF, PMC moves to KT and equilibrium is X^*.

If subsidy per product is paid by the government to a supplier the production level decreases, PMC becomes KT, because the decrease of one unit of $\hat{X}$ is PMC = GX + EF paid by the government. Thus, $\hat{X}$ also is optimal.

Fig. 2. OQ is the supply of claims for gas discharge at the optimal level set by the government. D is demand level by the producer of X who is also the discharger of gas. It demands claims instead removing CO_2 gas. Thus D expresses the MC level. The cost is the same to pay QM to the government or to pay QM as PMC for equipment. The government can control the optimal CO_2 level by supplying claims.

<u>Fig. 1</u>.

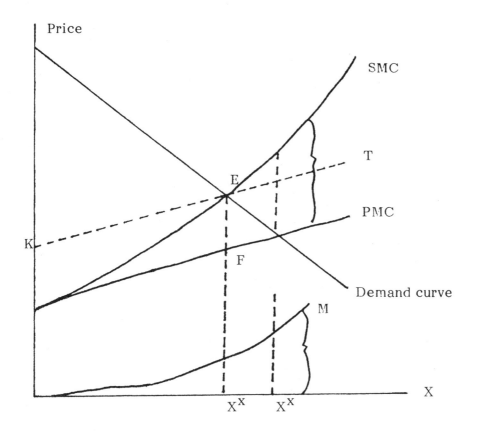

Fig. 2.

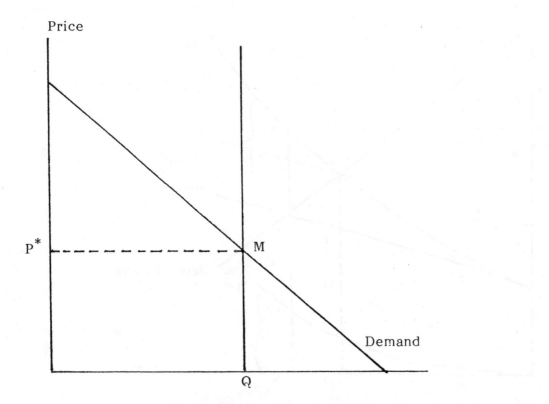

claims = volume of polluted gas distribution

Fig. 3. GDP per capita in terms of the ratio of heavy industry as a whole.

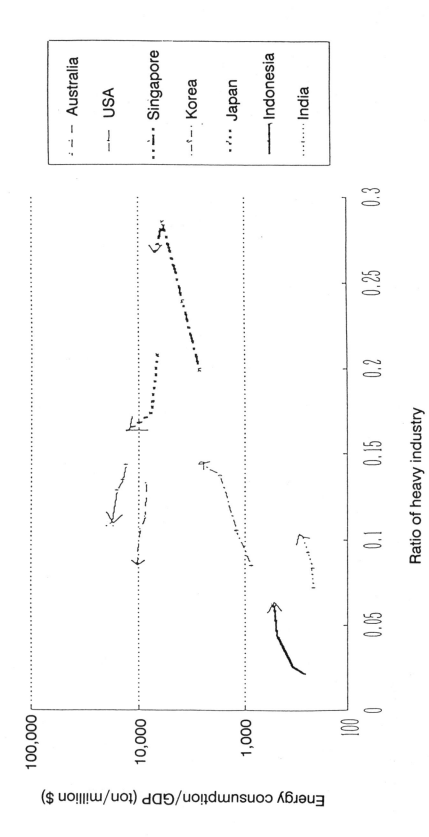

Fig. 4. Energy consumption as a function of GDP per capita and the ratio of heavy industry in industry as a whole.

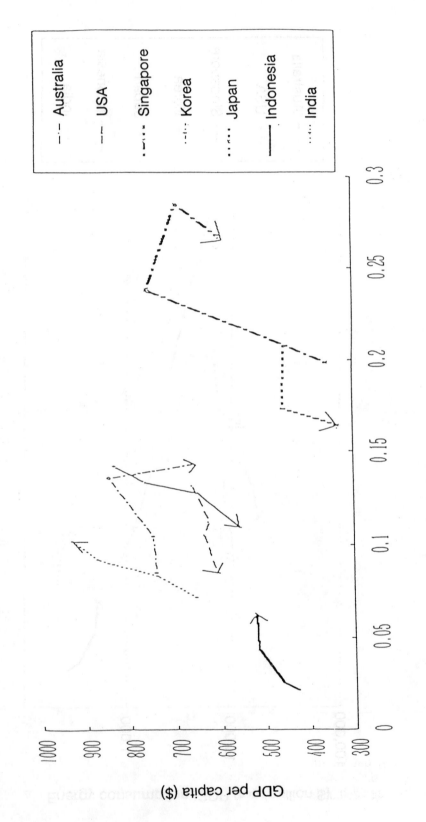

REGENERATION OF TROPICAL RAINFOREST ON NEW BRITAIN ISLAND, PAPUA NEW GUINEA

Seiji Mori
Project Coordinator
New Business
Nissho Iwai Corporation
Tokyo, Japan

1.   Introduction

It is expected that the world population will reach 10 billion by the middle of next century. This figure represents four times more people than in 1945. Most of the increase in population will occur in developing countries, creating such problems as shortages of food, housing, and fuel, and widespread poverty.

The present author worked in Papua New Guinea (PNG) for approximately ten years, from January 1976 to May 1985, as a managing director of the timber company SBLC located on New Britain Island. During the same period, I served with the Forest Industries Council, PHG, as an executive member and vice chairman for nine years. I proposed to the government in 1977 that foreign-owned timber companies should be responsible for carrying out reafforestation as a condition for the granting of timber-cutting permits, and this was registered under the new forest policy in 1979. This paper gives my ideas on reafforestation.

2.   General background on the tropical forest

Based on FAO data for 1981, the situation of the tropical forest and reafforestation efforts were as shown in **Table 1**. In 1990 the FAO warned that:

1)   the annual loss of tropical forest area was estimated at about 17 million ha, compared to 11 million ha in 1980; and

2)   the total area of tropical forest was estimated at about 1.7 billion ha, down from 1.9 billion ha in 1980.

The main reason for the decrease in tropical forest area is the increase in population, especially in developing countries.

3.   Reafforestation: Reasons, methods, and results

After World War II, logging operations changed remarkably, due to the development of heavy equipment. Once this equipment was applied to timber cutting in natural forest, conditions were rapidly degraded. In addition to deliberate felling, young trees of small trunk diameter and new shoots are felled by the machinery as well. At the same time, however, the construction of wide logging roads and better working conditions due to the use of heavy equipment have attracted more people to the timber industry, creating employment opportunities. Heavy equipment-related degradation of the forest has also resulted in another problem. Although previously tropical forest regeneration under natural growth conditions occurred in a 25-year cycle, it now requires 35 to 50 years for natural recovery.

Since SBLC has received an industrial land lease for 99 years in PNG, sustainable management of the tropical forest is in its own best interests. Depending solely on timber resources supplied by natural forest growth is not feasible for economic or environmental reasons. That would also increase log production costs due to the greater distances they must be transported in a depleted forest. Reafforestation helps to ensure and even increase a constant supply of commercial and industrial timber from the right species at reasonable cost.

Most commercial wood for plywood and sawn logs are dominant species in the tropical rainforest, including lauan (merantee), apiton (keruin), and capore. Ecologically, however, they grow below nondominant species which young. Nondominant species grow faster and have a shorter life cycle than dominant ones, and most are of inferior commercial quality. Since Japan imports a large quantity of tropical hardwood for plywood, there has been worry about the decreasing supply of resources. In order to obtain a steady, sustainable supply of large-diameter tropical round wood without knots for plywood use, the present author selected three nondominant species to plant in PNG: kamarere (*Eucalyptus deglupta*), erima (*Octomeles sumatrana*), and terminalia (*Terminalia brassii*). Their characteristics in the forest and at the sawmill and log yard were carefully studied, since neither erima nor terminalia had been planted in PNG at that time. In addition teak (*Tectona grandis*) was selected since it is used as a decorative species for veneer and quality furniture.

Table 2 shows the results of reafforestation using the four selected species. The reafforestation project was begun in 1975 and continued until 1978 in the first experiment with 255 ha of kamarere. Subsequently in 1981, the other three species were added. Sample plots for observation of growth and other factors were established in 1982.

The log volume harvested from the reafforestation area is ten times that of natural forest. Harvest time required for the nondominant reafforestation trees is approximately half that for the dominant species in the natural forest. Therefore, considering the increasing population and rising economic demand for logs, the government should carry out reafforestation on 10% of the forest where selective cutting has been done. The remaining 90% should be a natural growth area to protect wild botanical, animal, and insect species. This area could also be used for food and cash crop cultivation, depending on economic development in the future.

Based upon the growth results of reafforested areas described above (for tropical rainforest in a volcanic ash soil area), the price per log is estimated in **Table 3**. The figures are based on a reafforested area planted with kamarere (67.3% of the total area), erima (18%), terminalia (12.2%) (Tables 4-6), and teak (2.5%). The average FOB selling price is US$102.50/cubic metre and the average FOB cost is $54.72/cubic metre.

In terms of income for the timber company and government, if the reafforested and harvested area reached 1,000 ha/year the harvestable volume would be 300,000 cubic metres, costing $16,416,000 to produce (at $54.72/cubic metre), and selling for $30,750,000 (at $102.50/cubic metre). This represents a profit before tax of $14,334,000. If company tax were 35%, net profit would be $8,588,400. The reafforestation cost of $6,000,000 (at $6,000/ha) is to be reinvested. Thus the profit ratio before tax would be approximately 239% and after tax approximately 155%. Government revenues would amount to $13,358,100, consisting of company tax ($5,745,600), royalties ($3,000,000), and export levies ($4,612,500).

The above calculations are based on the following constraints on the financer(s):

1) Interest should be no more than 5% per annum on the principal.

2) Interest should be lent separately from the principal.

3) A grace period should be allowed for one year after the planted trees are harvested.

The government should provide the following guarantees to the national or foreign enterprise carrying out the reafforestation:

1) the right of land lease for reafforestation for long enough to harvest the commercial planted wood;

2) the right of ownership of planted trees; and

3) tax incentives against the cost of investment if the planter helps to set up processing facilities.

The benefits of such arrangements for developing countries are numerous, including:

1) more foreign currency income;

2) higher government revenues;

3) more employment opportunities;

4) technical transfer from developed countries;

5)   fewer environmental problems; and

6)   the availability of reserved land for future development.

4.   Conclusions

In October 1991, the present author attended a reafforestation workshop sponsored by the PNG government, and made the following suggestions. If PNG has 50 million ha of land, and 2% is volcanic ash soil area, there is 1 million ha available for the type of reafforestation project described above. If trees are replanted in a 20-year cutting cycle, the annual harvestable area will be about 50,000 ha, giving an estimated harvest volume of 15 million cubic metres per year, or almost the same annual export volume from Sarawak. The converted value of FOB log exports is expected to be about $1.5 billion, which is almost equal to the annual budget of the PNG government.

It was therefore recommended that the PNG government invite overseas investors to participate in reafforestation projects while giving guarantees for the security of investment. It is essential for the regeneration of the tropical forest that government, ODA agencies, and private enterprise work together. It is to be hoped that both developing and developed countries jointly promote reafforestation to resolve environmental problems while progressing with economic development.

Table 1. <u>Size of tropical by area, 1981.</u>

Area		Tropical America	Tropical Africa	Tropical Asia	Total
Area of tropical forest (1980) (unit: MIL. HA)	Closed Forest	679	217	306	1,201
	Bush forest	217	486	31	734
	Total	896	703	336	1,935
Annual decrease in area (unit 1,000 HA)	Closed forest	4,340	1,330	1,830	7,500
	Bush forest	1,270	2,340	190	3,800
	Total	5,610	3,670	2,020	11,300
Caused by agriculture		35%	70%	49%	49%
Manmade forest (unit: 1,000 HA)	Total area (1980)	4,620	1,780	5,110	11,510
	Annual planting (1981-1985)	530	130	440	1,100

Table 2.   Cutting cycle, yield, and sizes.

Species	Cutting cycle (yrs)	Standing tree Height (m)	D.B.H. (cm)	Standing volume (cubic metre/ha)	Harvested log vol. (cubic metre/ha)	Uses Plywood sawn	Pulp fuel
Eucalyptus deglupta (kamarere)	20	50-60	60-75	400-500	300-400	Yes	Yes
Octomeles sumatrana (erima)	15	40-50	80-85	400-500	300-400	Yes	Not suitable
Terminalia brassii (terminalia)	20	50	65-75	400-450	300-400	Yes	Not suitable
Tectona grandis (teak)	25	40	65-70	400	300	Yes	Yes

Table 3.   Estimated price/log (US$, cubic metre).

Species		Kamarere	Erima	Terminalia	Teak
1)	Cutting cycle (yrs)	20	15	20	30
2)	Yield of log (cubic metre/ha)	300	300	300	300
3)	Wood cost	$ 9.14	$ 7.85	$ 9.14	$ 11.70
a)	Planting	($ 800/ha)	($ 800/ha)	($ 800/ha)	($ 800/ha)
b)	Indirect cost	($ 400/ha)	($ 400/ha)	($ 400/ha)	($ 400/ha)
c)	Interest (8%) $72/yr	($ 1,440/ha)	($ 1,080/ha)	($ 1,440/ha)	($ 2,160/ha)
d)	Land lease ($5.0/ha/yr)	($ 100/ha)	($ 75/ha)	($ 100/ha)	($ 150/ha)
	(Subtotal)	($ 2,740/ha)	($ 2,355/ha)	($ 2,740/ha)	($ 3,510/ha)
4)	Royalty	$ 10	$ 10	$ 10	$ 10
5)	Logging-shipping	$ 20	$ 20	$ 20	$ 20
6)	Export levy	$ 15	$ 15	$ 15	$ 45
7)	Fob cost	$ 54.14	$ 52.85	$ 54.14	$ 86.70
8)	Fob sales	$ 100	$ 100	$ 100	$ 300
9)	Profit before tax	$ 45.86	$ 47.15	$ 45.86	$ 213.30
10)	Company tax	$ 16.05	$ 16.50	$ 16.05	$ 74.66
11)	Net profit	$ 29.81	$ 30.65	$ 29.81	$ 138.64

Table 4. **Growth conditions of kamarere (1990).**

Plot No.	Planted	Age	Locality	Topography	D.B.H. (cm)				Height (m)			
					Avg.	Top 100/ha	Balance	Top 80/ha	Avg.	Top 100/ha	Balance	Top 80/ha
2	1972	18	Mosa	Flat	41.6	49.4	28.9	51.9	43.7	49.4	34.4	50.8
3	1976	14	Buvussi	Flat	35.8	43.1	27.0	45.2	37.2	41.5	32.2	42.3
4	1976	14	Buvussi	Flat	37.4	44.2	29.6	45.9	38.5	41.7	34.8	42.0
5	1976	14	Buvussi	G/slope	45.1	-	-	45.5	40.6	-	-	41.5
20	1976	14	Buvussi	G/slope	44.1	45.8	30.8	47.5	41.5	44.2	34.8	45.7
6	1977	13	Buvussi	G/slope	35.8	43.5	28.2	45.2	39.3	44.0	34.7	44.8
7	1977	13	Buvussi	G/slope	33.8	39.3	26.9	41.0	37.8	39.3	26.9	41.3
11	1978	12	Malilimi	Flat	32.3	41.0	27.7	42.1	38.2	42.0	36.3	43.4
12	1978	12	Malilimi	Flat	31.3	39.8	26.6	40.9	35.7	40.3	33.1	41.2
13	1978	12	Malilimi	Flat	34.4	39.5	29.7	40.6	39.6	41.6	37.7	42.0
9	1982	8	Buvussi	Flat	27.2	35.2	22.1	36.3	29.7	33.5	27.2	33.8
10	1982	8	Malilimi	G/slope	25.0	34.5	21.7	35.4	28.9	33.2	27.5	33.5
8	1982	8	Buvussi	Flat	26.5	36.6	22.9	37.5	29.3	33.1	28.0	33.2
14	1982	8	Lakiemata	Flat	26.5	35.1	21.3	36.2	30.0	33.9	37.7	34.3
15	1982	8	Lakiemata	G/slope	28.4	35.2	21.3	36.3	29.8	33.7	25.7	33.8
16	1982	8	Lakiemata	Flat	24.8	32.1	19.5	33.0	27.1	30.9	24.4	31.0
17-A	1982	8	Lakiemata	Flat	22.6	33.8	18.3	34.7	26.8	33.5	24.2	33.6
17-B	1982	8	Lakiemata	Flat	29.4	35.1	24.4	36.0	29.9	33.2	27.1	33.2
18-A	1982	8	Lakiemata	Flat	22.7	31.4	19.0	32.3	26.8	33.6	23.9	34.3
18-B	1982	8	Lakiemata	G/slope	29.2	35.0	23.3	35.9	31.3	34.3	28.2	35.1
25	1984	6	Mopili	Flat	19.9	26.1	16.8	26.6	22.6	25.2	21.3	26.3
27	1984	6	Lakiemata	Flat	28.5	32.3	24.9	33.0	29.4	30.0	28.7	30.3
28	1984	6	Lakiemata	Flat	27.9	31.2	24.9	32.1	28.3	29.6	27.1	29.7

(Table 4. continued)

	Vol per stem (m3)			Vol per ha (m3)				Area of Plot	Existing stem per ha	Survival ratio (%)	Spacing
Avg.	Top 100/ha	Balance	Top 80/ha	Total	Top 100/ha	Balance	Top 80/ha				
2.737	3.828	0.978	4.255	443.52	382.86	60.66	340.40	0.5 HA	162 pcs	26	4 x 4
	(100)%	(86)	(14)		(77)						
1.800	2.643	0.797	2.935	331.32	264.34	66.98	234.80	0.5	184	29	4 x 4
	(100)	(86)	(20)		(71)						
1.970	2.800	1.027	3.049	370.40	280.02	90.38	234.92	0.5	188	30	3 x 3
	(100)	(76)	(24)		(66)						
2.780	-	-	2.963	244.70	-	-	237.00	0.5	88	14	3 x 3
	(100)				(97)						
2.514	3.081	1.097	3.376	352.02	308.14	43.88	270.08	0.5	140	22	4 x 4
	(100)	(88)	(12)		(77)						
1.810	2.719	0.919	2.969	365.68	271.92	93.74	237.52	0.5	202	32	4 x 4
	(100)	(74)	(26)		(65)						
1.554	2.045	0.788	2.236	254.99	204.56	50.43	178.88	0.5	164	26	3 x 3
	(100)	(80)	(20)		(70)						
1.392	2.297	0.915	2.438	403.58	229.70	173.88	195.04	0.5	290	46	4 x 4
	(100)	(57)	(43)		(48)						
1.253	2.067	0.796	2.205	348.47	206.74	141.73	176.40	0.5	278	25	3 x 3
	(100)	(62)	(38)		(53)						
1.546	2.083	1.050	2.200	321.76	208.36	113.40	176.00	0.5	208	33	4 x 4
	(100)	(65)	(35)		(55)						
0.852	1.422	0.475	1.514	214.59	142.28	72.31	121.12	0.5	252	30	4 x 4
	(100)	(67)	(33)		(57)						
0.703	1.350	0.477	1.432	272.65	135.08	137.57	114.56	0.5	388	35	3 x 3
	(100)	(50)	(50)		(42)						
0.810	1.513	0.559	1.586	307.98	151.38	156.60	126.88	0.5	380	61	4 x 4
	(100)	(49)	(51)		(41)						
0.824	1.416	0.476	1.516	219.21	141.66	77.55	121.28	0.5	266	43	4 x 4
	(100)	(65)	(35)		(55)						
0.945	1.418	0.452	1.514	185.35	141.96	43.39	121.12	0.5	196	31	4 x 4
	(100)	(77)	(23)		(65)						
0.666	1.104	0.349	1.174	158.158	110.42	48.16	93.92	0.5	238	38	4 x 4
	(100)	(70)	(30)		(59)						
0.595	1.070	0.333	1.359	216.80	128.88	87.92	108.72	0.25	364	58	4 x 4
	(100)	(59)	(41)		(50)						
0.969	1.387	0.609	1.458	209.30	138.76	70.64	116.64	0.25	216	35	4 x 4
	(100)	(66)	(34)		(56)						
0.573	1.113	0.344	1.190	192.76	111.36	81.46	95.20	0.25	336	30	3 x 3
	(100)	(58)	(42)		(49)						
0.997	1.434	0.560	1.524	199.52	143.42	56.00	121.92	0.25	200	18	3 x 3
	(100)	(72)	(28)		(61)						
0.363	0.625	0.232	0.654	108.98	62.58	46.40	52.32	0.5	300	48	4 x 4
	(100)	(57)	(43)		(48)						
0.849	1.088	0.626	1.141	176.51	108.82	67.69	91.28	0.25	208	33	4 x 4
	(100)	(62)	(38)		(52)						
0.786	1.003	0.591	1.057	166.71	100.38	66.33	84.56	0.25	202	19	3 x 3
	(100)	(68)	(40)		(51)						

Table 5.   Growth conditions of erima (1990).

Plot No.	Planted	Age	Locality	Topography	D.B.H. (cm)				Height (m)				Vol per stem (m³)				Vol per ha (m³)				Area of plot	Existing stem per	Survival ration (%)	Spacing (m)
					Avg.	Top 100/ha	Balance	Top 80/ha	Avg.	Top 100/ha	Balance	Top 80/ha	Avg.	Top 100/ha	Balance	Top 80/ha	Total	Top 100/ha	Balance	Top 80/ha				
19	1982	8	Lakimata	Flat	31.1	40.3	27.7	41.1	26.9	30.8	25.5	31.1	1.014	1.755	0.743	1.843	379.27 (100)%	175.56 (46)	203.71 (54)	147.44 (39)	0.5ha	374 pcs	89	5 x 5
26	1982	8	Lakiemata	G/slope	34.0	38.8	29.6	40.0	27.9	29.6	26.3	30.1	1.194	1.591	0.825	1.703	248.32 (100)	159.18 (64)	89.10 (36)	136.24 (55)	0.5	208	50	5 x 5
21	1983	7	Buvussi	G/slope	24.4	34.8	21.7	35.7	24.2	29.6	22.8	30.3	0.605	1.258	0.434	1.343	292.75 (100)	125.80 (43)	166.95 (51)	107.44 (37)	0.5	484	77	4 x 4
24	1984	6	Mopili	Flat	21.0	28.9	17.4	29.6	16.2	20.1	14.5	20.5	0.316	0.620	0.316	0.663	101.64 (100)	62.04 (61)	39.60 (39)	53.04 (52)	0.5	322	52	4 x 4

Table 6. Growth conditions of terminalia brassi (1990).

Plot No.	Planted	Age	Locality	Topography	D.B.H. (cm)				Height (m)				Vol per stem (m³)				Vol per ha (m³)				Area of plot	Existing stem per ha	Survival ration (%)	Spacing (m)
					Avg.	Top 100/ha	Balance	Top 80/ha	Avg.	Top 100/ha	Balance	Top 80/ha	Avg.	Top 100/ha	Balance	Top 80/ha	Total	Top 100/ha	Balance	Top 80/ha				
22	1983	7	Buvussi	G/slope	22.6	28.1	21.2	28.4	24.7	27.4	24.1	27.5	0.472	0.752	0.375	0.770	219.20 (100)	75.20 (34)	144.00 (66)	61.60 (28)	0.5	464	74	4 x 4
23	1983	7	Lakiemata	Flat	20.2	25.4	19.5	25.8	24.1	26.3	23.8	26.6	0.357	0.594	0.322	0.615	281.56 (100)	59.42 (21)	222.14 (79)	49.20 (17)	0.5	788	71	3 x 3
29	1983	7	Lakiemata	Flat	23.6	26.8	22.6	27.3	24.2	25.4	23.8	25.3	0.478	0.650	0.441	0.671	205.37 (100)	65.04 (32)	140.33 (68)	53.68 (26)	0.5	418	39	3 x 3

TROPICAL FOREST POTENTIALS, PROBLEMS, AND MANAGEMENT EFFORTS

Mohamad Soerjani
Centre for Research of Human
  Resources and the Environment
Environmental Science Postgraduate
  Programme
University of Indonesia
Jakarta, Indonesia

1.   Introduction

In their hypothesis on the emergence of the biotic components in a balance with the abiotic components of life on earth, Starr and Taggard (1984) stated that in appropriate conditions, spontaneous combination takes place between various simple molecules, which subsequently undergo stratification with the liposome layer (a kind of simple lipid chain). This thin layer serves as a membrane, permeable to water, but impermeable to ion, liquid, and elastic (self-repairing), which allows chemical substances to accumulate, thus swelling it until it breaks up into fragments. These fragments are in some aspects identical to their "parent." This, obviously, does not as yet imply "reproduction" in the real sense, but probably merely arbitrarily chemical "growth."

Kasting *et al.* (1988) have recently propounded a hypothesis concerning the decrease in the earth's atmospheric CO_2 content. This hypothesis was put forward in view of the assumption that the kind of life now existing on earth is only possible with quite a low CO_2 content in the atmosphere. This decrease in CO_2 is a geochemical process and has taken place from 500,000 years ago. CO_2 together with rainwater forms carbonate ($CO_2^{..}$), which reacts with rocks containing calcium-silicate (Ca-Si) producing carbonate rocks (CaO_3). Rainwater also releases calcium and bicarbonate (Ca^{++} and HCO_3) from carbonate rocks and groundwater. These ions are transported to rivers, lakes, and the sea where plankton, seaweed, shellfish, and other aquatic organisms, through a process of synthesis, fixate the $CaCO_3$ as a component of their bodies (shells, etc.). In this way CO_2 content in the atmosphere is decreased. Kasting *et al.* also explained in their hypothesis how through a process of subduction to the "bowels" of the earth, $CaCO_3$ is subjected to heat and pressure, resulting in a process of transformation which releases CO_2 back into the atmosphere. This happens through volcanic smoke and eruption, among others.

The Gaia hypothesis (earth's goddess) was put forward by Lovelock and Margulis in 1975 (Odum, 1983). This hypothesis stated that conditions on earth have developed under the influence of life in such a way that enables earth to support life. It is assumed that without life earth would have a very high atmospheric CO_2 content ($\pm 98\%$), whereas with life, which fixates most CO_2 (in the form of carbohydrates, fats, proteins, etc.), earth has only a low atmosphere CO_2 content (0.03% or 300 ppm). With such a low CO_2 content, earth is livable as it is now.

Obviously, the process diminishing CO_2 from 98% to 0.03% was a very long one, lasting for hundreds of thousand of years. However, after man has gone through a rapid sociocultural evolution backed by science and technology, it seems that things are again heading in the opposite direction. The relationship between earth that sustains life on the one hand, and the control of (man's) life in order so that it does not exceed earth's carrying capacity on the other, appears to be ignored (by man). Man as a "formidable reforming agent" has lost "basic wisdom," which particularly in this era of rapid modernization is badly needed.

For Indonesia, the forest in not only part of life but is, on the whole, its main support. The tropical forest in Indonesia covers 75% of the country's land area, or 50% of that of Southeast Asia, or 10% of the world's. It is of great and unique value for Indonesia and the whole world. Our ancestors' habitat was the forest, and for a number of our people it still is. Even as man's cultural revolution later brought into being artificial habitats, such as towns, villages, and settlements, with their ricefields, gardens, plantations, roads, bridges, etc., man's dependence on the forest was, and still is, obvious since the artificial habitats and all that goes with them are built on what used to be forest lands, using materials taken from the forest, and are on the whole unable to dispense completely with its support. We are at a stage of progress where our dependence on the forest is on the increase. The forest is the backbone of development. The main problem is that the capacity of the forest as a supporting agent for development tends to decrease, while our need for its support is rapidly increasing. Thus the balance and continuity of development are endangered.

This paper is of necessity based on a limited observation because of the simple fact that the problems related to the forest touch almost every aspect of life: development, economy, politics, security, loans, welfare, erosion, hydrology, greenhouse effect, store of living organisms, and esthetic values, ethics, to mention a few. Dealing with every problem involved and its intricacies is impossible. Added to this, the data available on the forest are very limited; most of the mystery is till beyond human ken. Even the data at our disposal are varied in the degree of veracity, differing in the numbers presented, thus giving rise to different arguments and conclusions. Therefore, this paper attempts to expose some differences in viewing the problems.

2.   The Tropical Forest

Life exists in different biomes, i.e., regional ecosystems with their corresponding biota. Seas, bodies of fresh water, and land are different biomes. Land biome comprises the tropical forest, the savanna, the temperate forest, the tundra, etc. The tropical forest is a biome characterized by an average temperature of about 25°C with slight diurnal, nocturnal, and seasonal variations. Humidity is 80% or more, and rainfall is generally high, over 400 mm annually.

The Indonesian tropical forest

Indonesia covers an area of 750 million hectares, 24.7% or 193 million hectares of which is land. The forest covers 75% or 144 million hectares of the land area. Tropical rain forest is mostly found in Kalimantan, Sumatra, Java, and Irian Jaya, while seasonal forest is found in East Java, Bali, West Nusa Tenggara, East Nusa Tenggara, Southeast Sulawesi, the Moluccas, and southern Irian Jaya. A smaller part of the land area is covered with freshwater swamp forest in the eastern part of Sumatra, South Kalimantan, and a little in Irian Jaya. Mangrove forest occurs in the eastern of part of Sumatra and the eastern part of Irian Jaya (see map in Fig. 1).

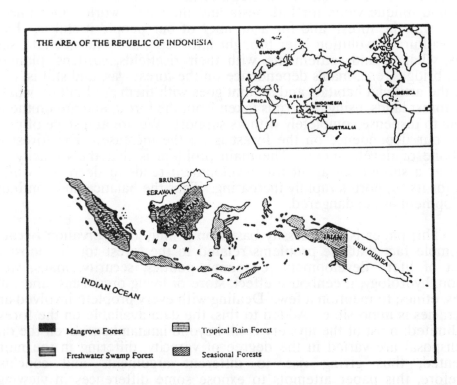

Fig. 1. **Map of distribution of tropical rain forest, seasonal forest, freshwater swamp forest, and mangrove forests (World Bank, 1988).**

In several areas, such as Muara Angke and Cilacap (around the Segara Anakan Inland Sea), mangrove forest also found, but not extensive enough to be shown on the map.

Land use in Indonesia

Of Indonesia's 193 million hectares of land area, 25.4% is a manmade environment, such as towns, villages, roads, industries, etc. The remainder, or 74.6%, is covered with conversion forest (15.8%), permanent production forest (17.6%), limited production forest (15.8%), conservation forest, national parks (9.7%), and protection forest (15.7%) (Fig. 2).

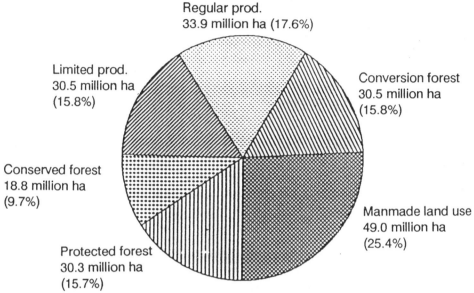

Fig. 2.   **Pattern of land use based on agreed forest areas.**

Figure 2 demonstrates that forest area still covers 74.6% of the land. If another 15.8% is converted into a manmade environment, the whole forest area will remain at 58.8%, while the manmade environment will increase to 41.2%. It should be noted that the 58.8% forest area is not all covered with vegetation. Part of it is in a critical condition; by 1989 almost 1.2 million hectares had been rehabilitated through reforestation, while in another portion of 5.8 million hectares rehabilitation/reforestation has **not** been carried out. For this reason and for the remaining 7.2 million ha waiting for regreening, the extent of forest area is only 109 million hectares, just 56.4% of the whole land area of Indonesia, and not 144 million hectares or 74.6% as mentioned above.

Tropical forest potential

The main functions of the forest in Indonesia are to serve as a biotic resource with its various biota, as a buffer of land and water and the earth's climate, and as a developmental back-up, as outlined below.

Indonesia is situated between two continents, Asia and Australia, and two oceans, the Indian and the Pacific Oceans. The influence of transition from Asia to Australia is particularly evident in the Wallacea area with the

Wallace line to the west marking the western border of the Australian biota distribution, and the Lydecker line to the east marking the eastern border of the Asian biota distribution. The stretch of area between the two lines is divided by the Weber line which marks the equilibrium border where biota are divided equally (50% each) between Asia and Australia (Fig. 3).

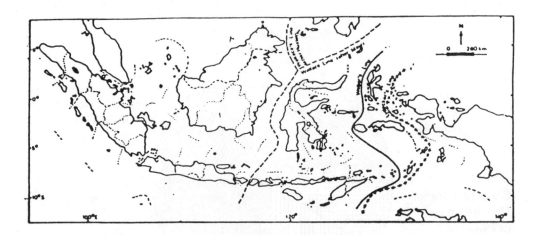

Fig. 3.   **The Wallacea area has the Wallace line as its eastern border, and the Lydecker line as its western. The Weber line is the biota equilibrium line when biota are 50% Asian and 50% Australian (see Soerjani, 1989).**

Due to the geographic situation, the forest in Indonesia is an unequalled storehouse of biotic resources. Of the living species existing on earth, more than 10% are found in Indonesia, of which 20% are vertebrate, 25% insects and reptiles, and more than 10% bacteria and fungi (see Table 1 for estimate of numbers of the main groups of the world's living species found in Indonesia).

Table 1. **Estimate of numbers of groups of species.**

Group	Global	Indonesia
Mammalia	40.000	3.000
Birds	8.900	1.500
Reptilia	8.000	2.000
Amphibia	6.000	1.000
Fish	38.000	8.500
Mollusca	150.000	20.000
Insects	1,250.000	250.000
Spermatophyta	300.000	25.000
Ferns	13.000	1.250
Lichens	16.000	1.500
Algae	21.000	1.800
Fungus	100.000	12.000
Bacteria/Cyanophyta	2.700	300

Source: Sastrapradja in Anon (1990a).

To preserve and develop biotic multiplicity, Indonesia has fixed a conservation area covering 17 million hectares consisting of nature reserves, fauna reserves, nature resorts, marine parks, and national parks scattered all over the country. These conservation efforts are mainly related to the function of the forest as a genetic resource storehouse. Genetic resources are the main factor in efforts to improve health, agriculture, animal husbandry, fishery, plantations, etc. The conservation areas include the following national parks: Mount Leuser, Kerinci Seblat, Southern Bukit Barisan, Ujung Kulon, Mounts Gede-Pangrango, Baluran, Western Bali, and Komodo. Nature resorts developed include Ir. H. Djuanda Forest Park in West Java, Dr. M. Hatta Forest Park in West Sumatra, and Bukit Barisan Forest Park in North Sumatra. To preserve and increase biotic multiplicity, various regulations and laws have been passed to protect endangered species, such as crocodiles, turtles, deer, orchids, etc. Endangered species of animals and plants needing protection number 521 and 36, respectively. The forest in Indonesia, owing to its biotic multiplicity, has a wide range of functions and useful potentials as shown in Table 2. The unique biotic multiplicity of forests which provides a wide range of functions is not sufficiently known, and we still tend to take short-cuts in the exploitation of forests.

The forest with its vegetation is closely linked with soil and water. Whatever happens to the forest will have effects on soil and hydro-orology. Conversely, the kind of soil and the pattern of hydro-orology will greatly affect the continued existence of the forest. For this reason, integrated forest management is tantamount to soil and water conservation. Indonesia has tens of river watersheds closely related to the problem of soil and hydro-orology conservation, where the forest is the main buffer. The 10 most important river watersheds in Java are: Ciliwung/Cisadane, Ciujung, Citarum, Cimahi, Citanduy, Serayu, Lusi, Serang, Solo, and Brantas.

Table 2. Functions and potentials of forests with biotic multiplicity.

Ecology	Direct use	Industry	Miscellaneous
1. Equilibrium preservation (ecosystem resilience: temperature, climate, biota)	1. Direct food (fruit, hunting results, sagu)	1. Timber Industry	1. Esthetics
2. Protection of life in nature	2. Material for drugs and food and beverages	2. Pharmaceutical industry (drugs, cosmetics, etc.)	2. Recreation
3. Protection of watersheds	3. Firewood	3. Paper industry (pulp)	3. Spiritual gain
4. Erosion control	4. Charcoal material	4. Tree liquid or sap (rubber)	4. Physical exercise
5. Water reserve storage	5. Construction timber	5. Residue (menthol, turpentine)	5. Love of nature
6. Absorbing CO_2 and other gases	6. Weaving material (fibres, silkworms)	6. Vegetative oils (clove, "adas," kayuputih)	6. History
7. Producers of O_2 other refreshing substances	7. Bee-keeping (honey)		7. Social cultural gain
8. Land fertility			8. National resilience

Areas with a 50% gradient are classified as critical, since the production potential is low, the protection function of hydro-ology is also insignificant, etc. By this definition, 7.8 million hectares or 60% of the island of Java is a critical area, including 1.9 million hectares of river watersheds of the upper reaches or 15% of the whole of Java, characterized by heavy land damage. Various rivers in Java have deteriorated seriously with the increase in mud content in the water, a situation running parallel to the increase in forest damage and the worsening of the hydro-ology of upstream river watersheds. The water of the Ciliwung River, for example, had a mud content of 1,150 mg/1 in 1911-1931, but had 36,500 mg/1 in 1974 or an increase of 32 times within 42 years. Another river with a similar rapid increase in mud content is Bengawan Solo. According to World Bank calculations (1990), the on-site cost of soil erosion in Java is estimated at US$315 million per year, while the off-site cost estimated at US$25-90 million per year. Loss of soil without conservation due to erosion can reach as much as 138.1 million tons/ha a year. Included in this calculation are the silting up of irrigation canals, the dredging of irrigation canals and harbours, and the loss in electric and irrigation capacity.

Despite the fact that Indonesia has a high rainfall, only about 50% of rainwater finds its way to rivers and because of the configuration of the river watershed only 35-45% of the surface water can be utilized. One of the causes is that the water current does not all come at the time it is needed. The 32 dams built in Java can only hold about 6.9 billion cubic metres or just 3.9% of all annual water current. The forest plays a very important part in increasing the stability of the current to the maximum.

The utilization of ground water is still very low in comparison with its potential. In Java, only 8.75 billion cubic metres of ground water is taken annually. Jakarta, which is the most densely populated area, uses the most ground water: 2 million cubic metres a day. This brings the problem of seawater intrusion which has penetrated as far as about 8 kilometres from the coast of Jakarta Bay. The possibility of the increasing use of ground water demands an increasingly greater role of the forest in supporting the aquifer which ensures the percolation process and rainwater penetration into the ground. The river watershed in the Jakarta area is in a very critical condition as the water catchment area in the south which used to be overgrown by forests, plantations, and garden vegetation has turned into a manmade environment with buildings, roads, and other structures, all of which are waterproof.

The forest has an important climatological function, particularly the absorption of CO_2 in the photosynthesis process, and simultaneously the release of O_2 in the same process. In its evolution, the earth is protected by gases called greenhouse gases, among which is CO_2 that sends off infrared rays, i.e., heat from sun rays reflected back by the earth's surface (see Fig. 4). In this way, the earth's temperature is kept warm, about 15°C on the average, and life on earth has gone through evolution adapting itself to the temperature. Without this greenhouse temperature, the earth's temperature would be as low as 18°C. If this should happen, life on earth would have to readapt. This would mean that a great number of species (probably including part of the human population) would not be able to cope with the new situation and perish.

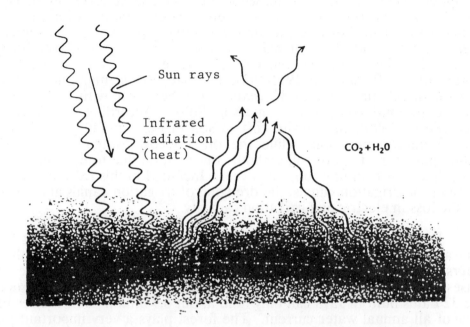

Fig. 4. **Greenhouse impact of gases, which reflects heat back to earth (from Kasting et al., 1988).**

The increase in man's activities on earth causes an increase in the greenhouse gas content made up of not only CO_2 but also of methane, CFC, etc. Methane originates from waste transformation, animal husbandry, agriculture, and swamps, while CFC (chlorofluorocarbon) is a gas used for cooling systems (refrigerators, air-conditioners), aerosols (perfumes, hair spray, pesticides), cleaning processes as of electronic instruments, production of foam, plastic glasses, plates, etc. CO_2 is a component of greenhouse gases originating from industry, transport, too much use of energy, logging, agriculture, etc. (Fig. 5).

Table 3.   <u>**Greenhouse gases.**</u>

Type of gas	% In atmosphere
CO_2	55
Methane	20
CFC	15
Other	10
Total	100

Annual increase in CO_2
in atmosphere

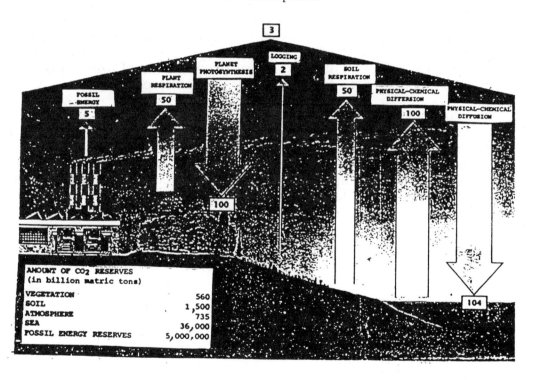

Fig. 5.   <u>**Annual balance of CO_2 in atmosphere. The figures show the annual addition and/or reduction of CO_2. Physiochemical processes in the sea release as much as 100 billion and absorb 104 billion metric tons of CO_2. Industry and logging release 2 to 5 billion metric tones of CO_2 to the atmosphere (Houghton and Woudwell, 1989; also in Soerjani, 1990).**</u>

From the illustration in Fig. 5, it seems clear that:

1)   The balance of physiochemical diffusion from the sea results in the reduction of 4 billion tons of CO_2 a year.

2)   The balance of vegetation respiration releases 50 metric tons of CO_2, while photosynthesis absorbs 100 metric tons a year, so that vegetation causes a reduction of 50 metric tons of CO_2 a year.

3)   The reduction of 50 metric tons CO_2 from vegetation balances with the release of 50 metric tons CO_2 a year from soil respiration (from waste, agriculture, decomposition of organic materials, etc.).

4)   The release of CO_2 from fossil energy, as much as 5 metric tons a year, is mainly from industries.

The total balance of the whole process and man's activities on earth add 3 metric tons of CO_2 a year. Further, as CO_2 is 55% of all of the greenhouse gases, the earth produces equivalent to 5.5 metric tons of CO_2 a year. We can conclude from this global figure the degree of importance of the forest as an "absorber" of existing CO_2. Even if this role could be intensified, the reduction of CO_2 sources (from the use of fossil fuels, logging activities) should be considered and controlled correspondingly.

Since Law No. 1, 1967, on the Role of Foreign Capital, and Law No. 6, 1968, on the Role of Domestic Capital in Forest Exploitation were enacted, numerous foreign and domestic companies have been active in the exploitation of forests. Also since then there has been rapid development that has increased the importance of the forestry sector in the acceleration of development in Indonesia. Forest products occupy the second rank after oil as a foreign exchange earner for the country (Table 4).

Table 4. **Export of forest products in 1984-1988 (in million US$).**

Product	1984	1985	1986	1987	1988	Average 1984-1988	%
Plywood	657,82	777,43	1.003,52	1.974,53	2.297,94	1.342,25	66.61
Sawn timber	305,24	334,64	438,96	632,62	713,09	484,91	24.06
Rattan	93,22	90,52	90,78	113,29	192,60	116,08	5.76
Other	212,83	61,46	52,07	42,46	89,70	91,70	3.56
Total	1,269,11	1,264,05	1,585,27	2,762,90	3,293,13	2,014,94	100.00

In addition, the forest sector employs about 300,000 workers directly, and about 700,000 indirectly. Up to the present, there has been an assumption the part of the general public that the forest only produces timber. According to estimates, wood consumption the world over amounted to 3,159 cubic metres in 1980, of which 54% was firewood while the remainder was used by industries (Fig. 6).

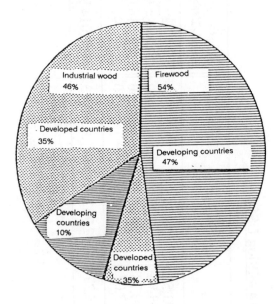

Fig. 6. **Global use of wood as firewood and industrial wood (Myers, 1985).**

The exploitation of the forest for its wood necessitates the felling of trees, which at the same time means the "crushing" of various other functions and potentials of the forest (Table 2). The amount of forest products other than wood and rattan, do not comprise more than 4.35%. These include "*tengkawang*," resin, turpentine, *kayuputih* oil, carbohydrates, etc. which could actually be more than 4.35%. Production could be stepped up of pharmaceutical materials for such medicines as morphine, codeine, quinine, atropine, etc, as well as the new possibilities for cancer medicines such as vinblastine, vincistrine, etc. which can be extracted from forest vegetation. It is worth noting that the American National Cancer Institute (NCI) is investigating the effectiveness of various vegetative extracts against 50 types of tumor cells including those of brain tumor and, leukemia, and as an antidote against the AIDS virus. The NCI is now collaborating with the Bogor Botanical Garden (Research Centre for Biological Research and Development) in the exploration of forest vegetation in Indonesia to find medicinal substances. The share of Indonesia as the owner of this forest resource and as a partner in this cooperation should be questioned and accounted for.

3.   Problems

This section deals with a few problems, i.e., the assumed deterioration of forests, logging, reafforestation, and the (direct) benefits people gain from forests.

Logging

Year after year the felling of forest trees is increasing worldwide. The percentage of logging relative to the forests owned is particularly on the increase in India, Costa Rica, Myanmar, Brazil, and Vietnam, while a more moderate increase is seen in Cameroon and Indonesia (Table 5). In a comparison with the average world logging percentage (1.4%), the histogram in Fig. 7 reveals that the percentage has been exceeded in seven countries. Cameroon and Indonesia are below the world average. From Fig. 8 it seems clear that the extent of logging in Indonesia increased from 600,000 hectares in the years 1981-1985 to 900,000 hectares in 1989. But the percentage relative to world logging decreased from 19.9% to 7.5%, while that of Brazil increased from 49.1% to 66.5%. Open forests in Brazil are more extensive, i.e., 157 million hectares (or 43.9%) of the total of 357.48 million hectares, whereas those in Indonesia cover only 3 million hectares (2.6%) of the total of 113.895 million hectares. According to data obtained from the Ministry of Forestry, Indonesia as a whole cut 1.2% of total owned forests (Table 6).

Table 5.   **Estimate of logging in "closed" forests in different countries, and how it compares with logging worldwide.**

	Country	Yearly logging 1981-1985	Recent logging 1989
1.	Costa Rica	4.0	7.6
2.	India	0.3	4.1
3.	Thailand	2.4	2.5
4.	Brazil	0.4	2.2
5.	Myanmar	0.3	2.1
6.	Vietnam	0.7	2.0
7.	The Philippines	1.0	1.5
8.	**Indonesia**	**0.5**	**0.8**
9.	Cameroon	0.4	0.6
World***			1.4

*   According to the FAO definition, a "closed" forest is one with a canopy covering 20% of the area, while an "open" forest means a mixture of forest and grassland with at least 10% of the area covered by a canopy of trees.

**   From various sources (mainly FAO).

***   A study based on WRI, IIED, & UNEP source (1989 & 1990).

****   Calculated on the basis of logging data according to Myers (1985).

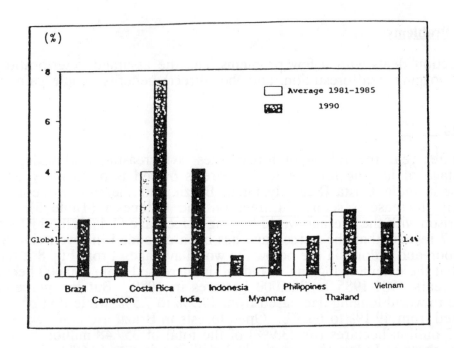

Fig. 7. <u>Logging percentage in nine countries in the years 1981-1985 and 1990, and the world logging average (1.4%) (adopted from WRI, IIED, & UNEP 1990).</u>

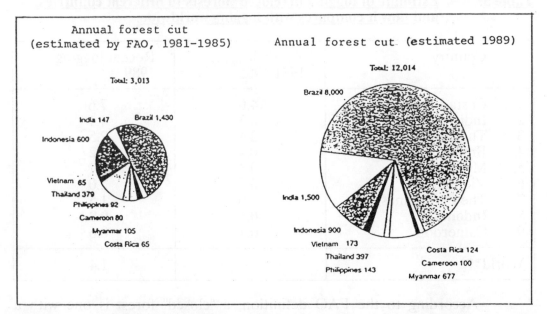

Fig. 8. <u>Logging in "closed" forests per year in various tropical countries in the years 1981-1985 and 1990 (figures are in 1,000 ha) (WRI, IIED, & UNEP 1990).</u>

Table 6. **Estimate of the amount of logging per area (1982-1990).**

AREA	Estimate of natural forest 1990 (ha)	Estimate of logging per year 1982-1990 (ha)	Percentage of logging per year (1990, in %)
1. Sumatra	20,382,400	367,700	1.8
2. Kalimantan*	34,732,400	610,900	1.6
3. Sulawesi	10,329,600	117,500	1.1
4. Maluku	6,029,300	24,300	0.4
5. Irian Jaya	300,649,000	163,700	0.5
6. Nusa Tenggara dan Timor-Timur	2,356,300	14,100	0.6
7. Bali	128,900	400	0.3
8. Java	965,400	16,100	1.7
Total	108,573,300	1,314,700	1.2

* Including the + 3 million ha burnt in 1982-1983.

Table 6 shows that logging in Indonesia is on the average still lower (1.2%) than the world average (1.4%). However, it is true that there are areas where the percentage is higher than that of the world's, namely in Java (1.7%), Sumatra (1.8%), and Kalimantan (1.6%). The high percentage for Java needs to be questioned. This probably includes the felling of "cultivated forests" which consist of such trees as teak, mahogany, albizzia, etc., and possibly also trees growing in people's gardens. Therefore, logging as related to the extent of the forest area does not seem to be the main problem. Attention should rather be paid to the manner of logging which may damage the land, and may be inimical to the continued existence of the biota in all its diversity. Above all the problem of how to rehabilitate damaged forest requires the most attention.

Reafforestation

While the forest has produced benefits for those exploiting it, these benefits have caused the deterioration of the forest in a number of aspects, such as its ecological function, its direct usefulness to common people, its nonwood products that are destroyed together with the trees, and its biodiversity. Even if reafforestation is effected, biodiversity can never be restored to its former state. Therefore, logging should be done intelligently, in order that biotic reserves, nature reserves, and the like can be kept intact. On paper all this has been well regulated, but in actual practice much remains to be done.

Reafforestation of logged forests and the greening of critical lands are absolutely essential. Figure 9 demonstrates that a climax forest that has been logged will absorb $_2$ for the process of decay and transformation, and conversely produce CO_2 and other gases such as methane and hydrocarbon. In the process of reafforestation, growing trees will cause CO_2 to be fixed through the process of an increasing photosynthesis, which culminates in secondary forest, and ultimately the climax will be regained. At this climax stage, CO_2 fixed through photosynthesis will be in balance with CO_2 released through respiration. Likewise, O_2 used in respiration will be in balance with O_2 released in the photosynthesis process.

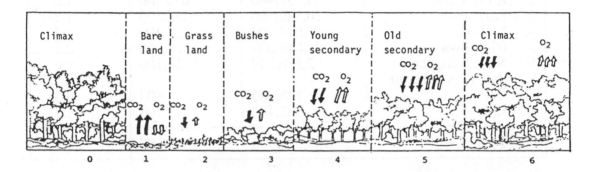

Fig. 9.   **The logged climax forest will release CO_2 and reduce O_2 for the process of transformation (1); during the successive stages in the process from grassland (2), bushes (3), to secondary forest (3 and 4), more CO_2 is absorbed in the photosynthesis process which releases more O_2 at the same time. When the climax has been reached, a balance between CO_2 absorption and O_2 release will recur.**

The important thing to note is that the period of time of successive stages until the climax is regained takes a long time, often up to 100 years or more. The problem, therefore, does not merely lie in the logging of the trees, but more in rehabilitation, reafforestation, and regreening, which have not been well dealt with in practice (Table 7). The Ministry of Forestry reported that in 1989 there was a discrepancy between the greening plan and the actual implementation covering 7.2 million hectares, while the figure for reafforestation was 5.8 million hectares. This fact demonstrates the discrepancy between activities for the sake of yield on the one hand and the price to be paid on the other. The culprits are partly logging and partly uncontrolled shifting cultivation. Particularly the slash-and-burn method and inappropriate land use in general, as well as malpractice in reafforestation, are to blame.

Table 7. **Rehabilitation of critical land: results achieved and unachieved target 1989.**

Area	Critical land rehabilitation programme (in hectares)			
	Greening		Reafforestation	
	Finished	Remainder	Finished	Remainder
1. Sumatra	1,323,003	2,298,600	493,580	1,405,900
2. Java	3,045,126	1,188,500		
3. Nusatenggara	468,811	1,225,900	124,654	1,034,500
4. Kalimantan	137,693	1,165,300	205,772	1,798,300
5. Sulawesi	835,016	965,200	395,893	1,099,300
6. Maluku	4,896	330,400	1,915	305,400
7. Irian Jaya		95,800		186,800
8. Lain Kepulauan				
Total	5,814,545	7,269,700	1,221,814	5,830,200

Benefits of the forest for the people

Since time immemorial people have regarded the forest as a source of their daily needs. For example, from the forest they get food (fruit, vegetables, animals), medicinal vegetation (quinine, kayuputih oil, herbs), refreshing substances (cinnamon, spices, menthol), firewood (branches, charcoal), construction timber and material for household utensils, materials for clothing and tools (silk, rubber), and many other benefits (including beekeeping). Then companies started to exploit the same forest products but on a large scale for commercial purposes. Unlike the common people, who take from the forest only what is necessary to sustain themselves and their families, companies take as much as possible for the market. In this era of rapid development, the need for land is increasing and in many cases forests have to be sacrificed. Likewise, the common people who have been living on whatever the forest can provide are also undergoing increased pressure. This has two important effects: the common people's source of living has become severely limited, and their attitude has changed in consequence. Those living on the fringes of the forest have had to abandon their traditional ways of interaction with the forest.

One example is shifting cultivation. In the past it was a wise, environmentally oriented, continuous activity. Now, because of the increasingly diminishing space in which to move, shifting cultivation has had to be abandoned. According to E.S. Thangam in Soerjani (1989a) the extent of forest land used for shifting cultivation is as shown in Table 8.

Table 8. **Extent of forest land used for shifting cultivation and the size of workforce involved.**

Data on	Extent (km^2)	Workforce
Forest land	1,439,542	-
Shifting cultivation activities	466,162 (32%)	-
Total population	-	17,136,113
Total shifting cultivators	-	2,740,615 (16%)

Source: Thangam (1989) in Soerjani (1989a).

Table 8 indicates that 32% of forest land is exploited by shifting cultivators who constitute only 16% of the whole population. This gap in percentage is accounted for by the fact that the existing workforce lack the necessary education and skill to enable them to benefit from natural resources, particularly the forest, efficiently and effectively. Moreover, the tools they use are so primitive that the yield is far below the maximum. The same applies to their physical condition. Their intake of nutrients and their environment in general are not conductive to good health. So shifting cultivation, in this age of scientific and technological progress, is an indication of: 1) dire poverty; 2) low educational level; and 3) poor health conditions.

For these reasons it is not the shifting cultivators alone who are to blame. It is the responsibility of all to better the situation. In general, the increasing benefits obtained from forest resources for development efforts should entail efforts for the improvement of the conditions of the common (poor) people.

4.   Forest Management with Environmental Considerations

Environmental Act

The Management of the Living Environment Act, known as UU (*Undang-Undang*), No. 4/1982, was enacted in recognition of the need to enhance effective management of environmental resources for the advancement of general welfare and happiness. This can only be achieved through sustaining the capability of a harmonious and balanced living environment to support continuous development, by means of an integrated and comprehensive national policy with due consideration given to the needs of present and future generations (Soerjani 1991; 1991a). In the general provisions of this act, the living environment is defined as a system comprising the natural environment, the manmade environment, and the social environment. These three components of the environment must be managed in balance, so that there will be harmony in their relationships and interdependence among the three components (Fig. 10).

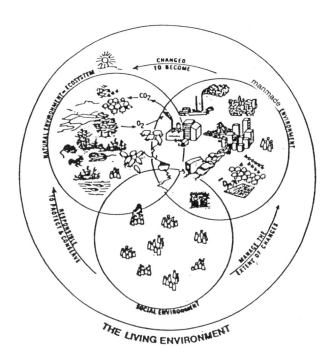

Fig. 10. <u>Concept of the living environment in which balance and harmony between the three components is the objective of development with environmental consideration (from Soerjani 1991).</u>

Present environmental concerns are largely due to the heavy pressure of development which rapidly increases the exploitation of resources, mostly using inappropriate technology, which causes deterioration in the quality of the environment. Under these conditions, the government's priority is to achieve balanced economic growth to improve the welfare of the people and the quality of the environment. To achieve this development is still considered the driving force, and therefore it is inevitable that developmental projects must be implemented. However, these must be assessed or evaluated to manage the detrimental impact upon the environment.

In response to the potential conflict between development and the environment, the 1982 Environment Management Act stipulates that the objectives of environmental management are to:

1)   achieve harmonious relations between man and the living environment with the objective being the development of the Indonesian individual in his totality;

2)   control wisely the utilization of natural resources;

3)   develop the Indonesian individual as a proponent of the living environment;

4)   implement development with environmental considerations, in the interest of present and future generations; and

5)   protect the nation against the impact of activities outside the state's territory which cause environmental damage and pollution.

The principles of environmental management are based upon sustainability and capability of a harmonious, balanced environment to support continued development for the improvement of human welfare. The act further stipulates that plans for all projects and activities that may create a significant environmental impact must include an **environmental impact analysis** document and **plans** for **environmental management** and **monitoring**. The philosophy behind this environmental management policy holds that the manner in which human actions are performed depends largely on the conditions and quality of environmental components. Therefore, environmental management should not be addressed toward managing environmental components as a strategy, but focussed more toward managing **human activities**.

Environmental Impact Analysis Regulation

In recognition of the fact that any activity including forest management may have an environmental impact, the Environmental Management Act was followed by Government Regulation (PP or *Peraturan Pemerintah*) No. 29 of 1986 concerning Analysis of the Impacts Upon the Environmental (EIA or AMDAL, *analisis mengenai dampak lingkungan*). The identified impact will be used as a basis for appropriate management of the environment and subsequent monitoring system. Following the EIA Regulation of 1986, any development project that falls into the following categories must be accompanied by and environmental information presentation (EIP = PIL or *penyajian informasi lingkungan*) before decisions concerning its approval will be made:

1)   processes and activities that may cause changes in the land structure and landscape;

2)   processes and activities that involve the exploitation of renewable and nonrenewable sources;

3)   processes and activities that can potentially cause the depletion, degradation, and deterioration of natural resources;

4)   processes and activities that may affect the social and cultural environment;

5)   processes and activities that can interfere with the protection of natural resources or the conservation of the country's cultural heritage;

6)   introduction of exotic plants, animals, and microorganisms;

7)   production and use of biotic and nonbiotic materials; and

8)   application of technology predicted to have great potential to affect the environment.

If the impact is considered important or significant, plans must be accompanied with an EIA document (ANDAL, *analisis dampak lingkungan*). Significant or important impacts are determined by:

1)   the number of people affected by project;

2)   the size of the impact area;

3)   the duration of the impact;

4)   the intensity of the impact;

5)   the number of components affected by the project;

6)   the accumulation of the impacts; and

7)   the reversibility or irreversibility of the impacts.

The fundamental principle behind EIA is that the desirability of development activities must not be based solely on their technological and economic feasibility, but also on environmental considerations. A project is only feasible if it meets technological, environmental, and socioeconomic feasibility requirements through a comprehensive screening process (Fig. 11).

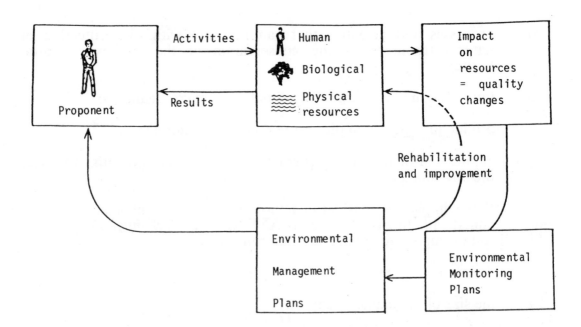

Fig. 12. **The impact of activities by a proponent which may create environmental impact on human, biological, as well as physical resources, to be followed by environmental management plan and environmental monitoring plans (Soerjani 1991a).**

Environmentally Sound Forest Management

One of the most important steps in forest management was the establishment of Central EIA (AMDAL) Commission of the Department of Forestry in 1988 by the Ministry of Forestry following the decree of the Minister of State for Population and the Environment No. KEP-53/MENKLH/6/1987. The commission is chaired by the Director General for Forest Protection and Nature Conservation and has members from the other institutions under the Ministry of Forestry as well as members from outside, including representatives of the Agency for Environmental Impact Management (BAPEDAL, *Badan Pengendalian Dampak Lingkungan* established in 1990 and chaired by the Minister of State for Population and Environment with two deputies for environmental pollution control and development), from the Ministry of Internal Affairs, other experts, and representatives from related ministries or institutions.

The central AMDAL Commission of the Department of Forestry issued AMDAL guidelines for forest development and exploitation in March 1989. The guidelines stipulate that those who plan and/or have any activity dealing with forest resources must submit an AMDAL document followed with plans for impact management and monitoring. The guidelines cover

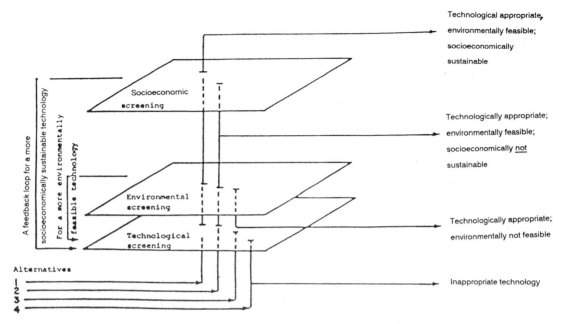

Fig. 11. Feasibility analyses or screening or projects. Project no. 1 would be approved for implementation, since it is technologically, environmentally, and economically feasible (modified from Beale 1980; Soerjani 1991 & 1991a).

It is expected that the three screening processes will be integrated into one feasibility study for a decision-making process. The document after being studied by the EIA Commission will be granted permission (or rejected) by the respective government institutions.

For projects that have been implemented or were in progress as of the date of implementation of the regulation (June 5, 1987), the project owner must prepare an evaluation concerning the impact upon the environment, consisting of one or both of the following document(s):

1) PEL (*penyajian evaluasi lingkungan*) or environmental evaluation presentation (EEP) if the project is considered to have minor impacts; or

2) SEL (*studi evaluasi lingkungan*) or environmental evaluation study (EES), a more thorough document, if the project is considered to have important environmental impacts.

The model in which the impacts are assessed for both the ongoing as well as proposed projects is shown in Fig. 12.

activities in the **preparatory phase** (including the potential survey), the **implementation of forest exploitation** (road construction, base camp, timber cruising, tree-felling method, transportation, log yard or log pond, rehabilitation (nursery, enrichment planting, regreening and reforestation, forest fire protection, conservation of water, soil, etc.). It must also include the end or **termination** of the right (report on the results of replanting, plans for further exploration and exploitation, etc.). The study must include the baseline of the **environmental profile** (climate, geomorphology, hydrology, soil, flora, the socioeconomic-cultural profile, etc.), the **potential impact** as well as the **evaluation** and **plans for the management** and **monitoring** of the impacts. Based on the recommendation of the commission, the official permit is granted by the respective institution.

Vast areas of forest land have been degraded and priority must be given to their rehabilitation. Although Indonesia is unable as yet and may never be able to dispense completely with forest exploitation, the extent of logging should reach a stage of stability, probably 1½ to 2 million hectares of forest a year. This should also include the loss caused by conversion, shifting cultivation, forest fires, and other sorts of loss. This stability should be reached in the sixth Five Year Development Plan in the year 2019. The target for reafforestation and regreening should cover 10 million hectares a year and should be achieved in the seventh Five Year Development Plan in 2043. This scheme should include urban forests, industrial forests, village social forests, etc.

The scheme for "industrial forest" or "forest plantation" is ambitious, but should nevertheless be started. In order that this scheme will be of direct benefit to the common people, village cooperatives, particularly in villages adjacent to forest, should be given the opportunity to take over (part) of the management of forest plantations. These plantations, which according to plan will cover 4.4 million hectares, will require more detailed preparations and dedicated, professional personnel. The production target is 9.7 million cubic metres per annum in the year 2000, which should increase to 2.5 million cubic metres a year in 2018. Timber is also produced in nonforest areas, such as rubber, coconut, and other plantations. The production forests should also be developed to produce nonwood products, particularly in the "agroforestry" pattern, which include resine, tanin, spices, essence oils, seeds and nuts (macadamia, cashew, nutmeg, etc.), and medicinal plants. The production target for nonwood products for the year 2018 is 25% of all the foreign exchange earned by forestry products, which is expected to increase to 50% in 2043.

Poor communities on the fringes of the forest depend on it for their living. This cannot be done away with. Unless opportunity is given to these communities to participate in the management of the forest, the function of the forest as a source of livelihood cannot be maintained. Production-sharing between them and companies seems to be the solution. For example, the communities or their cooperatives can own shares in state-owned companies exploiting the forest. In cases where the companies are wholly private, the system of intercropping and production-sharing of one sort or another can be applied. In any case, the communities should share the benefits of the forest with the government and large companies as environment-conscious forest exploitation develops.

In the second Indonesian Forestry Congress held in Jakarta in 1990, criticisms against logging received a great deal of attention. Indonesia became one of the targets of these criticisms for alleged uncontrolled logging, particularly from certain groups in Europe and the USA. They based their criticisms on the issue of global climate change. In some countries a boycott against timber from Indonesia was organized, especially after the wood industries in Indonesia managed to step up the production of sawn timber in 1985-1987 to 1½ times that of 1975-1977. There are also those who refute the criticism by saying that the antitropical wood campaign is actually prompted by the desire to stop the growth of Indonesia's timber industries, which might endanger or compete with similar industries in advanced countries which use imported tropical wood as raw materials. There are even groups who want to stop development in developing countries altogether, or at least hamper them in their efforts to "catch up" with advanced countries, by launching the issue of the irreparable harm done through logging the tropical forests. Thus there seem to be deliberate endeavours to perpetuate the gap between developing countries and advanced countries.

WRI, IIED, and UNEP (1989) suggest that a balanced analysis should be carried out of the deterioration of not only tropical but also of temperate forests. The data from 1989 indicate that forest cutting in North America to obtain logs, firewood, and industrial timber rose drastically in percentage, even higher than that in Indonesia. According to a 1986 data analysis, in Europe the damage done to conifer forests decreased, but the opposite happened to broad-leaf forests. Between 1978 and 1981, great forest fires raged in the Mediterranean, accounting for 83% of total forest fires in Europe. These fires destroyed 99% of the forests in the region. No data on forest fires in (the former) Soviet Union are available. When more complete world data have been gathered, better knowledge of the situation will permit meaningful efforts to alleviate problems.

A more rational attitude suggests that if logging the tropical forests is to be terminated, development that has been paid for by forest products must be subsidized by foreign governments. In concrete terms, it has been proposed that debts incurred by developing countries should be waived on the condition that they cease logging forests. There is another proposal urging that tropical forest be "bought" by the world. The "purchase," which should be defrayed by a world fund, would then provide the owning countries with enough compensation to maintain their balance of payments without having to depend on forest products.

The Netherlands is still willing to purchase timber from developing countries on the condition that the forests should be restored. No clarification is given as to the restoration in this case, which should be accompanied by such requirements as the enforcement of laws and "the rules of the game." Several institutions in the Netherlands such as the NEPP

(National Environmental Policy Plan) of the Ministry of Economic Affairs are involved in designing policies concerned with timber. It is planned that the use of hardwood originating from tropical forests is to be restricted from 1995. As compensation, developing countries concerned will be assisted in development projects.

Forest management and sustainable society

A set forth in the recently joint publication of the IUCN-UNEP and WWF (1991) **Caring for the Earth**, the strategy for future forest management in a sustainable society will include the following elements:

1) *Respect and care for the community of life.* This reflects the duty of care for other people and other forms of life, now and for the future. This includes care for the diversified forms of life in forest ecosystems.

2) *Improve the quality of human life.* This include processes to build self-confidence and allow people to lead lives of dignity and fulfilment. Life will be more meaningful if one can do more for others, including care for forest resources although it may not be directly related to one's own benefits.

3) *Conserve the earth's vitality and diversity.* If there is a need for forest exploitation and utilization of its resources, there must be assurance that the uses of reneweable resources are sustainable.

4) *Minimize the depletion of nonrenewable resources.* Resources (e.g., minerals and fossil fuels) that are located underneath a forest ecosystem must be conserved as much as possible.

5) *Stay within the earth's carrying capacity.* This must be adjusted to local conditions, since the limits vary by location and forest type.

6) *Change personal attitudes and practices.* To adopt the ethic for living sustainably, and the unavoidable dependence on nature and forest ecosystem, there must be a reexamination and readjustment of human values and human behaviour.

7) *Enable communities to care for their own environments.* This must include efforts to accelerate people's partnerships, ownerships, and responsibility to manage forest resources in a sustainable way.

8) *Provide a national framework for integrating development and conservation.* This includes the conservation of forest resources for sustainable support of development.

9) *Create a global alliance.* No nation today is self-sufficient. There is a need to establish a firm global alliance in forest management, if we are to achieve global sustainability.

Bibliography

Abelson, P. H. (ed.). 1984. *Biotechnology and Biotechnological Frontiers*. Boston, MA: Association for Advancement of Science.

Ministry of Forestry and FAO. 1990. *Indonesia National Forestry Action Plan*.

Anonymous. 1990a. *Population and the Living Environment*. Jakarta: Office of the Minister of State for Population and Environment.

Anonymous. 1990b. *Conclusions of the Second Indonesian Forestry Congress II*. Jakarta, 22-25 October, 1990.

Anonymous. 1991. Tropical rain forest. *Environmental News from the Netherlands*, no. 2: 3-6.

Houghton, R. A. and Woodwell, G. M. 1988. Global climatic change. *Scientific American*, 260: 18-27.

IUCN, UNEP, WWF. 1991. *Caring for the Earth*. A strategy for sustainable living. Switzerland: Gland.

Kasting, J. F., Toon, O. B. and Pollack, J. B. 1988. How climate evolved on the terrestrial planets. *Scientific American*, 258 (2): 46-53.

Myers, N. (ed.). 1985. *The Gaia Atlas of the Planet Management*. For today's caretaker of tomorrow's world. London: Pan Books Ltd.

Odum, E. P. 1983. *Basic Ecology*. New York: Holt & Saunders.

Soerjani, M. 1987. Some thought on the permanent agriculture in a sustainable development. *National Conference on Permanent Agriculture*. Jakarta: Ministry of Forestry.

Anonymous. 1991. Environmental Management Act and EIA regulation in Indonesia. *Conference on Environmental Regulation in Pacific Rim Nations*. American Bar Association & University of Hong Kong, Hong Kong, February 26-28, 1991.

Anonymous. 1991a. Environmental profile on Indonesia and the role of universities in promoting appropriate environmental management. *Seminar on the Role of ASAIHL Universities in Promoting Preservation of the Environment*, Airlangga University, Surabaya, May 6-8, 1991.

HYDROLOGICAL EFFECTS OF DEFORESTATION AND ALTERNATIVE LAND USES: A FRENCH EXPERIMENT IN THE AMAZONIAN RAIN FOREST

Jean-Marie Fritsch
Senior Hydrologist
French Institute for Scientific
   Research and Cooperation for
   Development (ORSTOM)
France

1.   Introduction

The experiment described in this paper took place in French Guyana, a country located on the Atlantic coast of South America, between the northern part of Brazil and Surinam (Fig. 1). French Guiana covers 90,000 km^2 and more than 90% of the country is under tropical rain forest, more precisely amazonian primary forest. This unusual situation accounts for the very small number of inhabitants living in the country (about 80,000 people), most living in the cities of Caiena (the capital) and Kourou or in smaller towns along the coast. This coastal strip is the area of economic activity and until the late 1970s nothing significant had been carried out inland. At that time, in order to bring economic development to the country, several huge projects involving US and French paper pulp companies were set up. The aims were to build up paper pulp factories, to exploit the rain forest for timber and paper pulp, and to plant fruit trees or grass for cattle breeding on the deforested areas or fast-growing trees such as pine and eucalyptus for the future supply of the pulp factories. The total area involved in the projects was around 1,000,000 hectares.

Although the authorities were interested in return on their investments, they were equally concerned about the negative effects of such large-scale deforestation by heavy machinery, and a multidisciplinary research programme was set up, taking into account all the aspects of the natural ecosystem and the impact of its transformation, known by all the acronym ECEREX (ECology, ERosion, EXperimentation).

This paper presents the results achieved on experimental watersheds in the fields of hydrology and erosion during this research programme. The experiment was conducted by ORSTOM (French Institute for Scientific Research for Development through Cooperation) and CTFT (Centre Technique Forestier Tropical).

2.   Methodology

The impact of the development projects on the water cycle and on erosion were specifically addressed through a study carried out on experimental watersheds. Ten small watersheds, with areas between 1 and 2 hectares were selected under natural forest conditions. To ensure accurate hydrometrical measurements, the control sections were equipped with 30° V-notch sharp-crested weirs for seven of the watersheds and H-flumes for three of them. Suspended sediment concentrations were measured in 2-litre water samples taken during the floods and the bed load was calculated from the volumes trapped in concrete sediment pits built upstream of each weir. The waters levels were monitored by high-speed chart recorders. A daily rain gauge recorder was installed in a clearing near each hydrometric station.

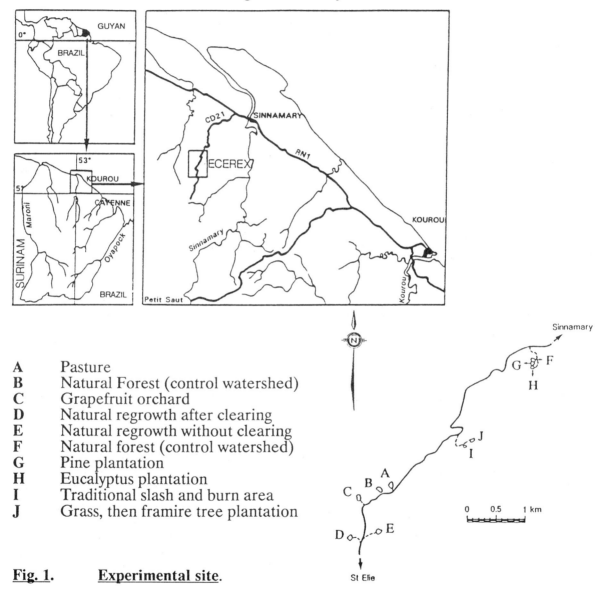

A	Pasture
B	Natural Forest (control watershed)
C	Grapefruit orchard
D	Natural regrowth after clearing
E	Natural regrowth without clearing
F	Natural forest (control watershed)
G	Pine plantation
H	Eucalyptus plantation
I	Traditional slash and burn area
J	Grass, then framire tree plantation

Fig. 1. **Experimental site**.

The watersheds are located on both sides of the St. Elie track (Fig. 1). They are identified by the letters A to J, following the order of the beginning of observations. Measurements began in January 1977 on A and B and in December 1979 on I and J. The distance between the two more distant watersheds, D and H, is 5 km.

The **paired watershed** approach (Hewlett and Helvey, 1970) was used to assess the effects of deforestation and alternative land uses on hydrological regimes. For all watersheds hydrological monitoring under natural forest conditions was carried out during at least a two-year period. The data collected during this **calibration period** were used to:

1)   assess the hydrological variability in space in natural conditions; and

2)   define and calibrate regressions of selected hydrological parameters on paired watersheds (quickflow, base flow, peak discharges, suspended sediment flows, etc.).

At the end of this **calibration period**, two basins assigned as **control watersheds** were kept in their natural condition, while several kinds of treatment were tested on the eight remaining **experimental watersheds**. During the **treatment period**, the effects were directly observed and measured on each experimental watershed and compared to the behaviour of this watershed as if it still had been under forest conditions. The natural behaviour was reconstructed with the data collected during the experimental period on the control basins, using the paired watershed correlations established during the calibration period.

3.   Characterization of experimental sites

Climate

As suggested by the latitude (5°30'N, 53°W) of the site, the climate is of the humid tropical type with the following characteristics.

Warm and **constant temperature around the year**.   The annual average temperature is 26.1°C, with very little seasonal variation (maximum 26.6°C in October, minimum 25.9°C in June). Diurnal variation is also relatively low, with maximum ranges of 6-8°C in the rainy season and 10-12°C in the dry season.

High rainfall, not evenly distributed around the year.   The average interannual rainfall on the experimental sites is **3,350 mm**, with extreme observed values of 2,394 mm and 3,680 mm. The rainy season lasts from December to June and there is relatively dry season from July to November. The wet season usually shows a trough in February or March (Fig. 2). The wettest month is May, with an average around 550 mm, but values over 1000 mm are not uncommon and have been observed twice during the last 10 years.

Although short-duration rainfall does not show exceptional values as tropical cyclones never hit the region, the one-hour precipitation nevertheless reaches 50 mm for a two-year return period and 70 mm for a 10-year return period. For maximum daily rainfall, the values are 145 mm and 200 mm, respectively, for the two-year and the 10-year return periods.

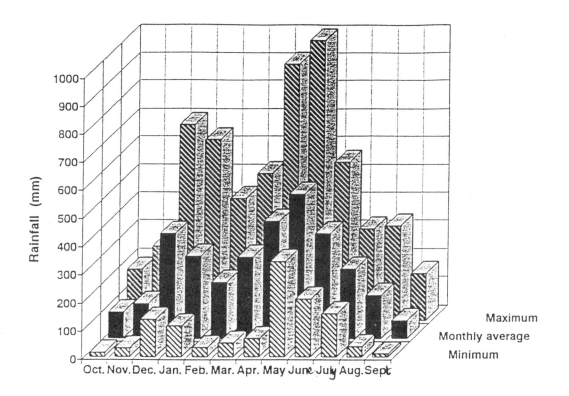

Fig. 2. **Monthly rainfall (mm) at ECEREX site (watershed A).**

Vegetation

The very large number of vegetal species is one of the most typical features of this forest: more than 200 species of "trees" (i.e., plants with stem diameter greater than 20 cm) were identified during an exhaustive inventory carried out on 42 hectares. Very large trees are not very numerous, on an average of 224 trees per hectare, and 168 stems were between 20 and 40 cm and only three were over 80 cm. However, on the other hand, the vertical development of the trees is extensive: 50-60 m is a common value for canopy height. The biomass was calculated by exhaustive sampling on 2,500 m^2 plots. Biomasses vary from 270 tons per hectare to 372 tons per hectare (dry weight). More than 90% of this weight comes from trees more than 20 cm in diameter.

Geology

The geological basement is composed of mica-schists. But these rocks only come to the surface in the form of deep alteration cover. The landscape is composed of series of small hills less than 100 m high, with steep slopes (15%

to 40%). As soon as the watersheds reach a few hectares in size (2 to 5), the streams come out onto flat valley bottoms, where runoff spreads out during the rainy season, making hydrometric measurements difficult and inaccurate.

Soils

The soils developed on the alteration cover belong to the red ferralitic soil family (French classification system) and the mineral and textural compositions are quite homogeneous all over the area. But as far as infiltration is concerned, and contrary to normal expectations, the soils show a very broad range of behaviour (Boulet, 1979). In some areas the drainage of the profile is good and water infiltrates fast and deep, even during heavy storms. It is only after continuously wet periods that watersheds developed on these **vertical drainage (VD)** soils show significant floods, in terms of volume and peak discharge. Some hills are entirely composed of soils of this type, but VD soils are more frequent on the tops. The VD soils are the most suitable areas in terms of agronomical potential.

In other areas, infiltration is blocked some 20-50 cm below the surface by a layer whose structure is relatively more compact than the upper one. Short-lived perched water tables and pockets of stagnant water are formed in the upper horizon. The soils are characterized by an internal **lateral drainage (LD)**, rather than by VD which becomes very poor. Even after moderate rainfall, strong superficial runoff is generated by processes known as saturation excess overland flow and return flow (Dunne, 1978). These soils are widespread and they make up the most common type in the region.

Another important hydro-pedological feature in this area is the occurrence of **water tables** in the bottoms. For some of the watersheds, the water table comes up to the surface during the core of the rainy season. These water tables are generally fed by lateral internal drainage along the slopes, but can be found as well in the bottoms of watersheds entirely composed of VD soils. The runoff generated by direct precipitation over the open water tables (100% of the incident rain) can be a very significant process in the runoff budget, as in some areas the extension of the water tables may reach up to 20% of the watershed area.

The combination of these three features (**VD** soils, **LD** soils, and water table occurrence) results in very different hydrological regimes on watersheds that nevertheless have similar features in area, shape, slope, and vegetation.

Table 1. Characteristics of the experimental and control basins (area, slope, soils).

Basin characteristics	C	I	E	D	B	A	J	G	F	H
Drainage area (ha)	1.6	1.1	1.6	1.4	1.6	1.3	1.4	1.5	1.4	1.0
Slopes % (maximum on each bank)	20-17	23-23	30-20	28-18	17-17	20-20	32-29	34-26	35-31	24-19
VD soils area (%)	99	60	57	60	10	0	2	0	0	0
Water table area (%) (coming to surface during the rainy season)	0	0	0	0	0	0	0	10	4	14

The main characteristics of the experimental watersheds are given in Table 1. One can see that the **VD** soil type ranges from 0 to 99% and that three basins frequently have water tables coming to the surface, with typical extensions between 4% and 14%.

4.   Hydrological regimes under natural forest conditions

On such small watersheds, the major part of the runoff occurs in the form of stormflow, as base flows usually cease after several days without any rainfall. This is the prevalent situation for seven basins; only three of them (F, G, and H) have a very weak continuous flow during dry periods. Quickflow was separated from total stormflow by the recession curve technique (Dubreuil, 1974), but interpretations of the results given by this method, as with any other conventional graphic technique, are fairly reliable. Taking into account the small size of the drainage basins, the analysis will concentrate on the variations of **total stormflow** volumes in natural and man-made conditions.

Hydrological variability in space

A two year period (1978-1979) with all the basins in natural forest conditions was available to assess the variability of runoff in space among the set of the experimental basins (Table 2). As shown in Fig. 3, **storm runoff ranged from 1 to 5**; more precisely, stormflow volumes varied from 7.3% to 34.4% of the rainfall.

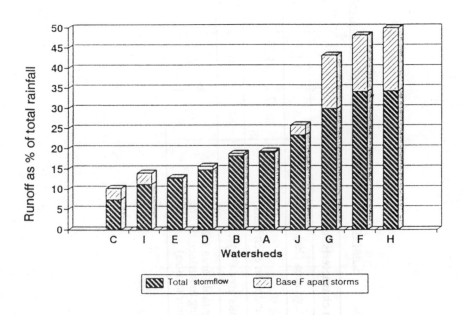

Fig. 3.   Variability of runoff for the ECEREX watersheds under forest (1978-1979).

Table 2.   **ECEREX watersheds in natural forest conditions. Rainfall and runoff. Interannual averages (1978-1979) (all values in mm).**

Watershed	A	B	C	D	E	F	G	H	I	J
Rainfall	3,423	3,267	3,265	3,257	3,350	3,102	3,173	3,165	3,285	3,219
Stormflow % of rain	650 19.0	595 18.2	239 7.3	480 14.8	426 12.7	1,058 34.1	947 29.9	1,088 34.4	364 11.1	748 23.3
Total flow % of rain	665 19.4	615 18.8	332 10.2	511 15.7	434 13.0	1,493 48.1	1,370 43.2	1,577 49.8	460 14.0	831 25.8

As previously mentioned, the soil conditions are mostly responsible for these variations, shown by the regression between the percentages of soils with **VD** and stormflow volumes (Table 3 and Fig. 4).

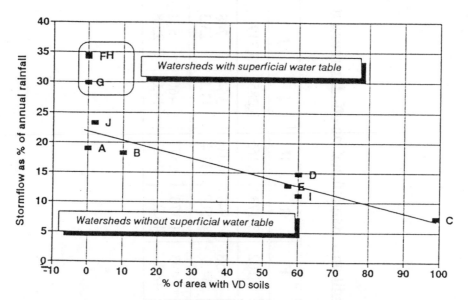

Fig. 4. <u>Storm runoff variability and soil types (interannual values in forest conditions).</u>

Hydrological variability in time

Runoff variability in time in the natural ecosystem can be assessed through the control watersheds that remained untouched during the experiment. Data collected on control watershed B during seven years show storm runoff variations between 300 mm and 723 mm, i.e., **a range of 2.4** (Table 4 and Fig. 5).

Table 3. Variability in space and soil types under natural forest.

Basin	C	I	E	D	B	A	J	G	F	H
Annual storm runoff % of rainfall	7.3	11.1	12.7	14.8	18.2	19.0	23.3	29.9	34.1	34.4
VD soils (% of area)	100	60	57	60	10	0	2	0	0	0
Water table extension* % of drainage area		-	-	-	-	-	-	10	4	14

* The *water table extension* was determined during the soil survey (the more frequent extension of the water table during the rainy season).

Table 4. **Variability in time (annual runoff of control watershed B under rain forest).**

Year	Storm runoff (mm)	Total surface runoff (mm)
1977	723	743
1978	541	563
1979	652	668
1980	437	461
1981	300	322
1982	485	505
1983	373	490

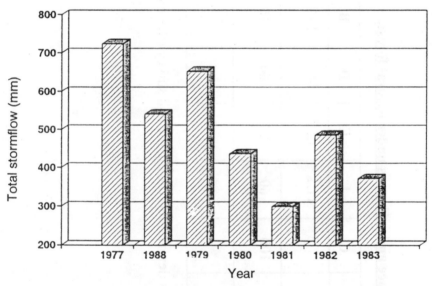

Fig. 5. **Interannual variability of storm runoff (controls watershed B, forest conditions).**

5.   Methods

Two watersheds were selected as control basins. Watershed F with strong runoff was assigned as the control for the basins with rising water tables and **LD** soil types (G, H, and J), while control B was used for the other basins where the water table dynamic plays a minor role and with mixed soil types (**VD** and **LD** types). This is the typical situation for basins A, D, E, and I. One experimental basin (C) with a very low runoff coefficient due to a high percentage of **VD** soils (99%) was monitored using the B control catchment as well, even if the hydrological processes and behaviour of the two catchments are significantly different.

The treatments applied were those that had been planned in the development projects and which are usually implemented in the region. **Logging** of large trees is the first step, which comprises following:

1) All trees more than 40 cm diameter are cut down with chainsaws. Crowns are cut. Stems are cut in logs.

2) A light caterpillar tractor (D4) equipped with a straight blade opens skidder access tracks (typical extension 240 m/ha).

3) The logs are yarded uphill by a rubber-tyred skidder.

At the end of this phase the area is neither clear-cut nor clean. The smaller trees are still uncut, all the roots are in place, and crowns and slash are left on the site.

Land clearing for agricultural purposes generally follows the logging. Clearing is achieved with heavy caterpillar tractors (D8 or D9) equipped at the front with a cutting blade (ROME or KG type) to fell the remaining trees, and at the rear with hydraulic claw system to pull out the roots. Finally, the slash (roots, stumps, small branches, leaves, etc.) is gathered along the contour lines by a caterpillar tractor with a rake blade (FLECO type). The slash is burnt whenever a relatively dry period occurs. Land clearing is a very detrimental practice for soil conservation, especially on steep slopes or when carried out during the rainy season.

Different scenarios, considered to be likely in the economic context, have been tested (Fig. 6):

1) The exploitation of the ecosystem is limited to the logging phase and no following action is taken. This is the **natural regrowing** scenario. Two different regrowing trials were tested, one in basin E with **natural regrowing** after logging only, and the other in basin D with **natural regrowing** after logging **and** land clearing.

2) **Plantation of fast-growing trees** after logging and land clearing was carried out with two species of trees. In basin G plantation of **pine trees** (*Pinus Caraïbea, var. Hondurensis*) and in basin H plantation of **eucalyptus** (*E. grandifolia, Flores*).

3) **Plantation of fruit trees.** The catchment with the best soil conditions (basin C) was dedicated to an orchard trial with plantation of **grapefruit trees** (pink pomelo).

4) Two trials were made with fodder grasses for grazing: in basin A **plantation of** Digitaria swazilandensis grazed by cattle and in basin J

plantation of <u>Bracharia decumbens</u>. This trial was later converted into a framiré tree plantation (*Terminalia ivorensis*).

5) **Traditional shifting cultivation.** Although this manual agricultural technique was not in the scope of the development projects, it was tested as a reference in basin I using **traditional slash-and-burn shifting cultivation.**

Six watersheds among the eight treated has completely bare soil at the beginning of the treatment (the exceptions were basin I converted in slash-and-burn cultivation and basin E on which all big trees were logged, but subsequent land clearing had not been done). This situation gave an opportunity for an assessment of the hydrological effects in a similar situation in terms of soil cover on watersheds which have been found to be very different in their behaviour during the calibration period under forest.

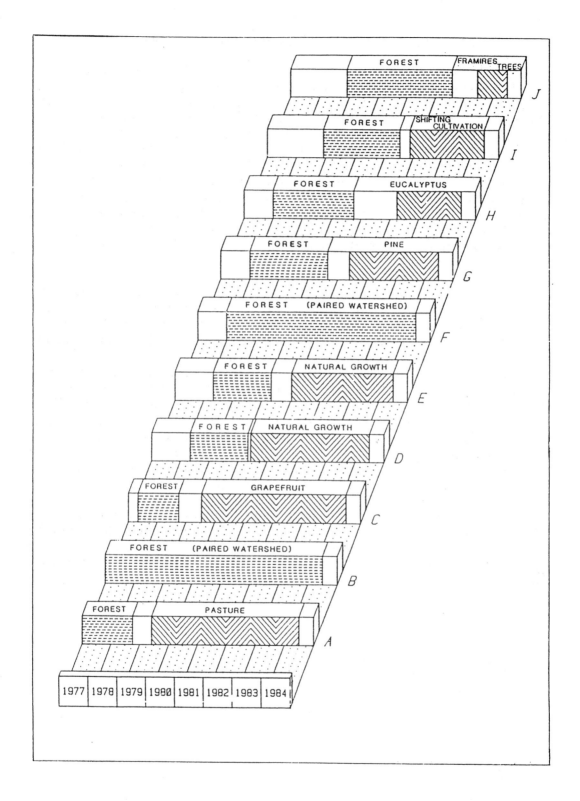

Fig. 6. <u>Synopsis of land use in ECEREX watershed experiment.</u>

6.   Methods for calculation of runoff in forest conditions

The double-mass curve is a very effective way to detect any break-up in trends, although accurate trend analysis can be problematic and may require some subjective interpretation. The double-mass curve was one of the techniques used in this study to detect and access the modifications on treated watersheds. Figures 7 and 8 show examples of mass curves of accumulated stormflow during the calibration period and during the logging, land clearing, and bare soil periods (control basin data are on the X-axis and experimental watershed data on the Y-axis). Figure 7 applies to A_B and Fig. 8 G_F paired watersheds. From now the notation X_Y is used for mass curves or cross-correlations of any parameter of experimental watershed X with control watershed Y. In these two cases, the effects of the treatments are neat due to similar soil conditions and consequent similar regimes of control and treated watersheds.

Correlation of stormflow volumes between control and experimental watersheds during the calibration period is another way to calculate the runoff a deforested watershed would have had if it were still in forest conditions. Several types of correlation were tested. The most reliable model was generally a linear correlation with runoff of the control as the first parameter and the difference of rainfall on the two watersheds as the second parameter. For some watersheds nonlinear correlations had to be used (i.e., for watershed C). Figure 9 is an example of the reconstruction of runoff for individual storms for watershed D, using a correlation calibrated on the pair D_B. Calculated values are on the X-axis and observed values on the Y-axis. Figure 10 shows the reconstruction of runoff for periods of 10 days for the pair J_B. The most significant correlation (for individual storms or over periods of 10 days) was used to access the effect of the treatments.

The accuracy of the calculation of the annual storm runoff within a 90% interval of confidence is given in Table 5. Runoff is generally calculated with an accuracy around or better than 5%, i.e., the method is able to detect in a significant way any modification in runoff greater than these accuracy thresholds.

Table 5.   **Accuracy of the reconstruction of runoff in forest conditions.**

Watershed	Accuracy for calculation of annual storm runoff (90% confidence level interval)
A_B	7.4%
C_B	12.0%
D_B	5.3%
E_B	5.6%
G_F	2.6%
H_F	3.2%
I_B	6.4%
J_B	4.3%

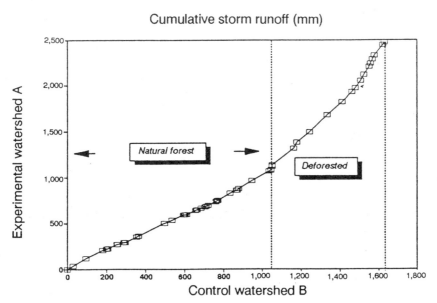

Fig. 7. Double-mass curve of storm runoff. Experimental watershed A and control B.

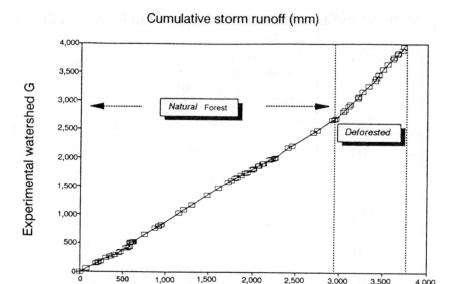

Cumulative storm runoff (mm)

Fig. 8. Double-mass curve of storm runoff. Experimental watershed G and F.

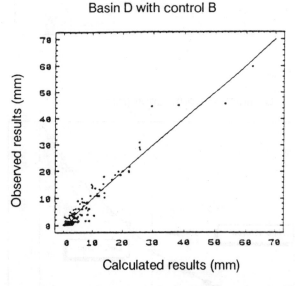

Basin D with control B

Fig. 9. Reconstruction of runoff by correlation for individual storms of watershed D.

Basin J with control B

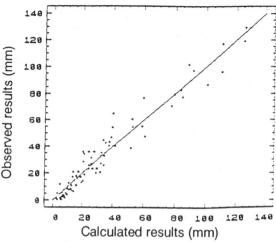

Fig. 10.   **Reconstruction of runoff by correlation for a 10-day period for watershed J.**

7.   Modification of runoff after land clearing (bare soil conditions)

Logging and land clearing were usually done during the dry season, except for sites G and H, where it was carried out at the beginning of the rainy season in a attempt to create the worst conditions reaching the upper limits of the impacts. According to each watershed, the bare soil period refers to different years, but always includes the main part of the rainy season, from January to July. Storm runoff increases (i.e., the part in the observed runoff due to treatment, expressed in mm) during this first rainy season after logging and clearing are given in Table 6.

Table 6.   **Increase in runoff during the rainy season following logging and land clearing.**

Watershed	Year	Observed runoff (mm)	Increase due to clearing (mm)
C	1979	682	304
D	1981	479	244
A	1979	1,616	762
J	1983	1,037	384
G	1981	1,388	621
H	1981	1,453	560

Some values are very high, and in fact are greater than the highest values quoted in the literature, for example 662 mm for Coweeta watershed, n°17 (Swank and Douglass, 1974) or 650 mm on a small watershed in New Zealand (bookman)

Zealand (Pearce *et al.*, 1980). These values of increase were obtained by the correlation method using the prediction of the mean. Figure 11 graphically shows these results as well as the values of the increases for the 90% interval of confidence and the increases calculated by slope analysis of the double-mass curves. *This clearly shows that the increases in runoff are high and also statistically significant. There is no doubt about the fact that mechanized forest logging followed by mechanical land clearing considerably increases the storm runoff of headwater streams.*

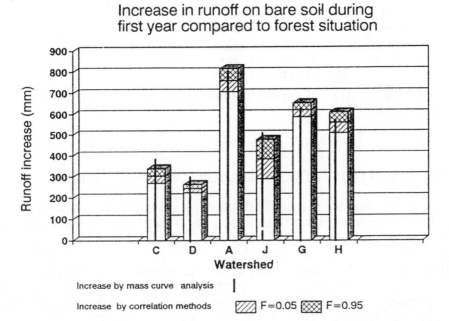

Fig. 11. <u>**Increases in runoff (in mm) during the rainy season following logging and land clearing. Calculated by correlation (for the mean and the 90% interval of confidence) and by slope analysis of double-mass curves.**</u>

However, these absolute values of runoff do not give a clear understanding of the effects of land clearing. For example, the data do not apply to the same year, so effects of interannual hydrological variability are included in the impacts of logging and clearing. In addition, during the first rainy season after clearing, it appeared that the analytical relationship between treated and control watersheds were unsteady during some weeks at the beginning of the period. This was due to slight differences and heterogeneity in treatments among the watersheds, and some time was necessary to come to achieve a stabilized response. The selection of the period of stabilized response was done by the statistical method of Bois

(1987). The results of the analysis carried out on the data of this steady behaviour period are given in Table 7 and are plotted in Fig. 12.

The ratios of increase after clearing, expressed as a percentage of the runoff of the respective watershed under forest, *ranges from 166% to 299%.* When the watersheds are classified in ascending order of their observed runoff during the calibration period under forest (i.e., C, D, A, J, G, H) the percentage of increase is in descending order. The lower the runoff, the higher the increase (Fig. 12); when expressed *in relative terms (%), increases are higher for the watersheds with low runoff in natural conditions rather than for those which had previously shown strong runoff,* i.e., the impacts of clearing are relatively stronger on better soil. However, *in absolute figures (mm), the highest increases were observed on the watershed that already had large stormflow runoff in natural conditions* (see dark shadowed areas in Fig. 13).

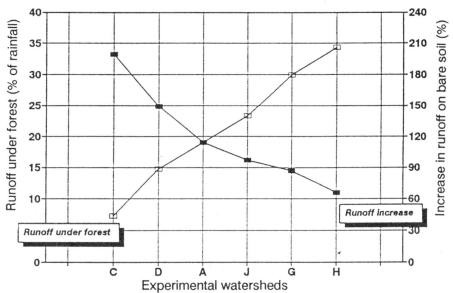

Fig. 12.   **Increase in storm runoff after clearing (%), compared with runoff under forest (%).**

Table 7. Increase in stormflow runoff after clearing (stabilized response period).

Watershed	C	D	A	J	G	H
Calibration period						
Observed storm runoff under forest (% of rainfall for 2 years period)	7.3	14.8	19.0	23.3	29.9	34.4
Treatment periods						
Rainfall (mm)	1,448	2,207	2,349	2,071	1,445	1,620
Observed storm runoff *"bare soil"* (mm) (% of rainfall)	342 23.6	450 20.4	1341 57.1	954 46.1	772 53.4	787 48.6
Calculated storm runoff *"forest"* (mm) (% of rainfall)	114 7.9	181 8.2	627 26.7	483 23.3	414 28.7	475 29.3
Increase in runoff with *bare soil* (mm) (% of rainfall)	228 15.7	269 12.2	714 30.4	471 22.7	358 24.8	312 19.3
Increase ratio of runoff after clearing (%)	299	249	214	197	187	166

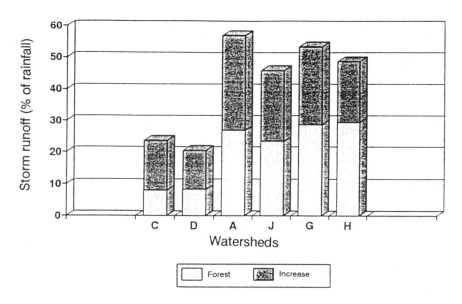

Fig. 13. <u>**Increase in storm runoff due to clearing (as % of rainfall).**</u>

8. <u>Evolution of runoff on treated watersheds</u>

The evaluation of runoff increases during the years following treatment was carried out using the methodology already described for the bare soil state (analysis of mass curve slopes, correlations for individual storms and for 10-day periods). The effects of treatments on the runoff on the experimental watersheds during the years following logging are summarized in Table 8. These results are presented graphically in Fig. 14 for experiments linked with forestry (tree plantation or natural regrowing) and in Fig. 15 for those that focus more on agricultural practices (pasture, orchard, slash and burn). One general conclusion is that runoff for which the increase had been very high during the year immediately following logging and clearing was reduced significantly on all watersheds. The results of more detailed analysis are described in the following subsections.

Table 8. Increases in stormflow runoff during the years following forest cutting.

	Year 1 Bare soil(*)	Year 2	Year 3	Year 4	Year 5	Year 6
C Grapefruit after clearing	199	73	17	63	46	
I Traditional slash and burn		23	30			
E Logging and regrowth without clearing		4	26	2	-6	
D Logging and regrowth after clearing	149	40	32	16		
A Grazed pasture after clearing	114	59	63	47	27	
J Grass plantation after clearing	97					
G Pine trees after clearing	87	62	33			-12
H Eucalyptus after clearing	66	47	12			-8

*: Stabilized response period.

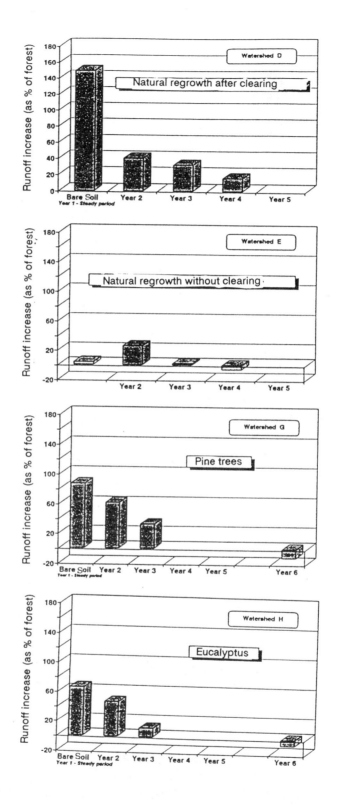

Fig. 14. **Variation in runoff in time for experiments linked with forestry (tree plantation or natural regrowth), expressed as a relative increase (in %) of the runoff in natural forest.**

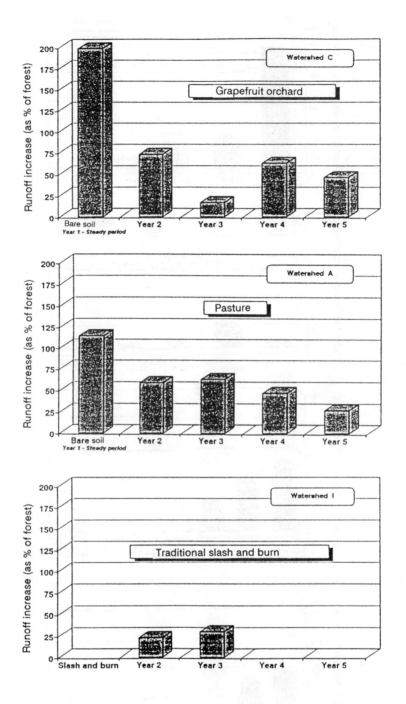

Fig. 15. **Variation in runoff in time for the experiments linked with agricultural practices (pasture, orchard, slash and burn), expressed as a relative increase (in %) of the runoff in natural forest.**

Differences between logging only and logging combined with land clearing

An interesting comparison can be drawn between the experiments carried out on watershed D (natural regrowth after logging followed by total land clearing) and watershed E (natural regrowth after logging only). As these two watersheds have shown very similar hydrological regimes in natural conditions (with rather low runoff coefficients, see Fig. 3) and were deforested during the same year, their respective variations of runoff can be directly compared, and we can assume it is the difference in treatment which accounts for the discrepancy between them.

Evidently, the severe logging taking all large trees (over 40 cm diameter) as was carried out on watershed E led to a maximum annual increase in runoff which was at most 26% higher than it would have been under natural forest, while with the additional effects off land clearing the increase was as high as 149% during the period of steady response. During this period (January to July) the rainfall was 2,207 mm with an observed runoff during storms of 450 mm, while the predicted runoff under forest would have been only 181 mm. *The hydrological effects of land clearing are extremely strong, even if compared with severe mechanized logging.*

During the fourth year after logging the runoff was only 16% greater than it would have been under forest for the cleared watershed. For the logged-only basin, the calculated increase is -6%, which means that runoff should have actually been less than it was in natural forest. In Table 6, the accuracy of the model used for reconstruction of runoff in forest conditions is not better than 5.6% within a 90% confidence interval. That means that the -6% value is not significantly different from zero and we must assume that the watershed has returned to its natural hydrological response as far as annual stormflow is concerned. Obviously, this return to figures similar to the natural ones does not mean that the ecosystem in its fullest sense has completely recovered its natural characteristics. As proved by the studies carried out in other disciplines, major changes will affect, among others, the botanical features (species distribution, biomass, size of trees) and the soil conditions (destruction of the upper organic horizon, etc.) for decades or even longer periods.

Fast-growing tree plantations

Watersheds G and H, which have the poorest soil conditions and consequently had the strongest runoff under forest, were assigned to the tests of fast-growing tree species, i.e., pine trees on G and eucalyptus on H. Although the increases in the first year were the lowest proportionally when compared to the other watersheds, the negative effects remained at a constant high level during year 2 after clearing. The figures were +62% (year 2) and +87% (year 1) for G and +47% (year 2) and +66% (year 1). It was only during year 3 after clearing that a significant reduction of runoff was observed on the eucalyptus plantation, while the value still remained at a rather high level of +33% for the pine plantation. The explanation for these figures can be found in the forestry management techniques: it is necessary to protect young pines and eucalyptus from natural regrowth whose development is faster than those of the plantations and therefore the area around them has to be cleaned regularly. During years 2 and 3 these forestry techniques artificially maintained soil conditions which were quite similar to those observed during year 1 immediately after land clearing, and accordingly the hydrological response was still high.

On these two watersheds, additional measurements were taken during year 6 after deforestation (i.e., on five-year-old trees), during a particularly dry year for that region as rainfall during the considered hydrological year (October 1986 to September 1987) was "only" 2,394 mm. These exceptional conditions made the construction of the natural forest runoff only slightly reliable as it was not possible to calibrate the regression model for such conditions. The conclusions are that during that particular year the runoff would have been less for the plantations than under forest, more precisely 12% for the pines and 8% for the eucalyptus. These conclusions should be taken as provisional and require additional measurements to confirm them.

Grazing experiment

This trial on a _D. swazilandensis_ grass plantation grazed by five to 10 young bulls per hectare (which is a load equivalent to 1,200 to 3,300 kg per hectare). Such a semiintensive design differs from the common ranching system in use in the amazonian region, with loads ranging from 0.28 animal per hectare (Fearnside, 1979) to 1.3 animal per hectare (Myers, 1982). Propagation by cutting was the technique used for plantation of the _D. swazilandensis_ grass, thus creating conditions for fast expansion of the forage and close vegetal cover a few weeks after the beginning of the rainy season. Nevertheless the runoff maintained high levels of increase for three years, with values around 60% for the first two years and 50% for the third year. It was only four years after plantation that a decreasing trend was observed with increases slightly under 30%. Thus, _grazing is an agroeconomic speculation which induces stormflow volumes definitely higher than under forest._ This conclusion is of some importance as grazing is widespread in tropical South American countries and stretches over large areas as the only form of land use.

Traditional slash-and-burn cultivation

No machinery was used on this watershed. Planting and harvesting were done by hand, the only mechanical input being the use of portable chainsaws to fell trees. This traditional cultivation is for family supply and comprises a broad variety of plants such as watermelon, corn, cassava, banana, pineapple, sweet potato, etc. Compared to most of the mechanized trials, the hydrological impacts were rather low. A 23% increase in runoff was observed during the first year and a 30% increase during the second. An extension of the area planted with corn during the second year may account for this difference. If these increases were actually slight, especially for a watershed where **VD** soils are dominant, _it cannot be assumed that manual slash-and-burn cultivation does not have any negative effect on the water cycle._ The effects are imperceptible on large watersheds, as long as this traditional system can remain a shifting cultivation system, which is only possible when low population pressure is applied on large areas. But with strong, continuous human pressure on large areas where slash-and-burn cultivation becomes dominant, the storm runoff will be increased in similar proportions as those observed in the experiment (+20% to +30%) during the first years, with likely increases during the following years as soil conditions get worse, leading to a reduction in agricultural productivity and to a subsequent deterioration in the quality of the vegetal soil covering.

Grapefruit orchard

Watershed C, with the best soils in terms of internal drainage and agricultural potential, was used for the plantation of grapefruit trees (480 small trees planted in a 7 x 5 m design). In addition, a grass cover of _Brachiaria USDA_ was planted between the trees to protect the soil. It was previously stated that watersheds with large parts of **VD** soils as for watershed C (where this percentage is 100%) are very sensitive to any type of treatment and that relative increases in runoff easily reach high values.

After an increase of nearly 200% calculated for the core of the first year rainy season for a bare soil situation, the increases dropped to lower figures such as 73% (year 2), 17% (year 3), 63% (year 4), and 46% (year 5). As the forest storm runoff of this watershed was very weak (i.e., 7.3% of the rainfall), the volumes of water corresponding to these data are nevertheless small, i.e., 196 mm (year 2), 25 mm (year 3), 141 mm (year 4), and 128 mm (year 5).

Looking for variation of increases in time could lead one to assume that the watershed shows uneven behaviour or that the data are unreliable (the figures decrease strongly between year 2 and year 3, after which runoff increases again). A point of methodology must be raised: it was indicated that the hydrological regimes of this watershed C and control B were very different, and so the reconstruction of the runoff of this experimental watershed as if it had been under forest was the least accurate estimation among the whole set of experimental basins. Figure 16 shows that this accuracy was not better than 12% (within a 90% confidence interval, which is not very severe). So, if the increases are calculated not only for the prediction of the mean but also for the limits of this 90% confidence interval, no significant differences can be identified between values characterizing years 2, 4, and 5, and not even between years 2 and 5 (Fig. 16). This shows that hydrologists must be very careful when assessing the effects of human activity on watersheds, because even if the effective hydrological runoff during treatment is measured in a fully reliable way, the final tools for the assessment will be calculation (mass curves, correlations or more complex mathematical models) which is never perfect and whose accuracy will determine the real capacity of assessing land use modifications. If the accuracy of the reconstruction model is x%, it will not be possible to detect modifications lower than x% in a meaningful way.

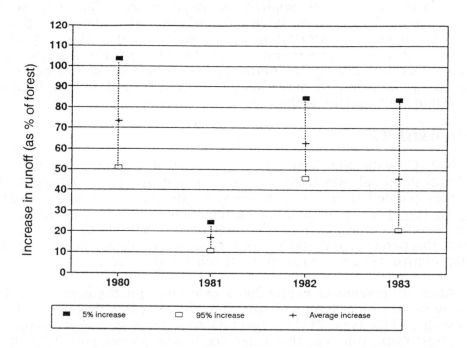

Watershed C - grapefruit

Fig. 16.   <u>**Range of calculated increase of runoff on watershed C.
Calculation of runoff increase for the mean and for the limits
of the 90% confidence interval of the reconstructed runoff
under forest.**</u>

9.   <u>Conclusions</u>

The information gained in the ECEREX watershed experiment can be
summarized as follows.

1)   As far as hydrological regimes of small watersheds are concerned, *the
natural ecosystem is very heterogeneous* in the area, a feature which was
not obvious when starting the studies. The *hydrological response to the
same amount of rainfall ranged from 1 to 5 in the natural ecosystem,
while the effects of treatments never exceeded a range of 1 to 3.* In
consequence, a precise knowledge of the initial situation is required
before any treatment is applied in order to assess exactly the effects of
land use changes. In this case the combined bidisciplinary approach
made by a hydrologist and a soil scientist has provided the basics for
such understanding.

2)   For six of the watersheds, on which mechanized logging and clearing
was conducted, a comparative bare soil situation was realized and a
steady response to rainfall was observed after some weeks. *Very high
increases in storm runoff were highlighted with averages over the core of
the rainy season ranging from +66% to +200% when compared with*

runoff under forest. The strongest impacts were observed on the watersheds initially having weak storm runoff and a general rule confirmed on all watersheds was that the lower the forest runoff, the higher the increase.

3)   Cutting the natural forest to change to any other land use among those tested will undoubtedly lead to an increase in quickflow runoff, a feature which must be considered seriously in all water management projects. Even for smooth alternative technology like traditional slash-and-burn cultivation, the increase in annual runoff can be as high as 30%. If grazing options are taken into account, these figures are much higher and may reach a steady level of +50% for several years. *It can be assumed that with all options except grazing the runoff will decrease and nearly return to the previous values after some years.*

4)   Research carried out by other scientists (working in soil and forestry sciences) has shown that the *hydrological behaviour cannot be used by itself as a marker of ecosystem conditions*. Runoff may have returned to figures close to the original ones, while the features of the regrowing forest and soils are still far from their original situation.

Many other results have been achieved (Fritsch, 1990) in the area of soil conservation (measurement of soil erosion and suspended sediment flow) and in the area of hydrology (effect of deforestation on peak flow, assessment of the effects of land use not only on the annual time scale, but also for individual storms and for 10-day periods). Figure 17 is an example of this continuous monitoring of the runoff variations for 10-day periods, for the grazing experiment (upper) and for the pine plantation (lower).

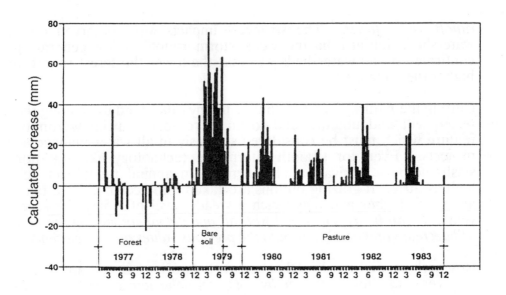

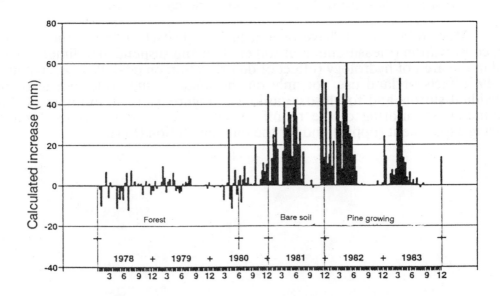

Fig. 17. Variations in storm runoff for 10-day periods on watershed A (pasture, upper panel) and on watershed G (pine plantation, lower panel). The bars show the calculated increase in runoff (in mm) for 10-day periods in comparison with forest conditions.

References

Bois Ph. (1987) Contrôle des séries chronologiques corrélées par étude du cumul des résidus. In: *Deuxièmes Journées Hydrologiques de l'ORSTOM*, Col. Colloques et séminaires. Paris: ORSTOM; 89-99.

Boulet R. (1979) Méthodes d'analyse et représentation des couvertures pédologiques des bassins-versants ÉCEREX. In: *l'Ecosystème Forestier Guyanais* Bulletin de liaison du groupe de travail. Cayenne: ORSTOM; n°1, février 1979, 11-17.

Dubreuil, P. D. (1974) Initiation à l'Analyse Hydrologique. Paris: Masson.

Dunne, T. (1978) Field studies of hillslope flow processes. In: Kirby M. J., ed. *Hillslope Hydrology* New York: John Wiley; 227-293.

Fearnside P. M. (1979) O desenvolvimento da Floresta Amazonica: problemas prioritàrios para a formulaçào de directizes. *Acta Amazonica* 9(4) suplemento: 123-129.

Fritsch J. M. (1990) Les effets du défrichement de la forêt amazonienne et de la mise en culture sur l'hydrologie de petits bassins versants. Doctorate dissertation, Université des Sciences et Techniques de Montpellier, November 1990.

Hewlett, J. D. and Helvey, J. D. (1970) Effects of forest clear-felling on storm hydrograph. *Water Resource, Res.* 6: 768-782.

Myers, N. (1982) Depletion of tropical moist forests: a comparative review of rates and causes in the three main regions. *Acta Amazonica* 12(4): 745-758.

Pearce, A. J., Rowe, L. K., and O'Loughlin, C. L. (1980) Effects of clearfelling and slash-burning on water yield and storm hydrograph in evergreen mixed forests, Western new Zealand. *Proceedings of the Helsinki Symposium*, June 1980, Pub. IAHS n°130, 119-127.

Swank, W. T. and Douglass, J. E. (1974) Streamflow greatly reduced by converting deciduous hardwood stands to pine. *Science* 185: 857-859.

II. Country Papers

COUNTRY PAPER: REPUBLIC OF CHINA

Kuo-Ching Chie
Senior Specialist
Council of Agriculture
 Executive Yuan
Taipei, Taiwan

Summary

As a result of heightened public awareness of the environment the present forestry policy and its implementation in Taiwan, ROC, emphasizes the long-term sustainability of all forest resources such as water, soil, fish, wildlife, and recreational sites. To fulfil this policy a multiple-objective management plan that considers both the conservation of forest resources and traditional needs should be carefully developed and effectively implemented.

1.   Introduction

Taiwan is a 3,577,000 ha island province of the Republic of China, lying off the southeast coast of mainland China to the north of the Philippines. It is neither large in size nor rich in natural resources, but the hard-working people of Taiwan, under the guidance of a progressive government, have built the island from postwar scratch into one of the most beautiful and productive places in the world. Some call its performance an "economic miracle" of recent times. Others consider it a successful showcase of free economic development.

The forests of Taiwan cover 1,865,141 ha, about 52% of the land area, with as many as 650 tree species. Since the forests of the world cover only 30% of the land area, Taiwan is a land of forests. Because of the small land area and dense population (20,514,477), the forest area per capita is only 0.1 ha., one-tenth of the world average.

2.   Forest Resources

Forest land area

About 52% (1,865,141 ha) of the land area of Taiwan is covered by forests. Three-fourths of the forest land is in national forests managed by the Taiwan Forestry Bureau. Most of the remainder is also national forest land managed by various other governmental organizations. The other public and private forests occupy only about 12% of the total forest.

It is clear that the national forests contain the majority of Taiwan's forest resources. Accordingly, effective management of national forests is a crucial responsibility for forestry agencies to assure the well-being of the whole nation.

Forest growing stock

The growing stock of Taiwan's forest was estimated to be 326,421,000 cubic metres. Most of these timber resources are: hardwoods 45%, coniferous 38%, and mixed conifer-hardwoods 17% (Table 1).

Table 1. **Forest resources of Taiwan, ROC (units: Area: ha, trees: 1,000 m³, bamboo: 1,000 culm).**

Forest type	Area	%	Growing stock	%
Hardwoods	1,157,927	62.0	146,877	45.0
Coniferous	416,700	22.5	124,550	38.0
Mixed forests	157,500	8.8	54,994	17.0
Bamboo	133,014	7.0	1,168,713	
Total	1,865,141	100.0	326,421	100.0

Source: <u>Forestry Statistics of Taiwan</u>, Taiwan Forestry Bureau, 1991.

Protected Forest

According to the Forest Law, the establishment of protected forests is an effort to prevent flood and typhoon damage, to secure water resources, and to protect lives and property. More protected forests are being set aside under the Forest Law in order to conform to changing needs and the condition of the environment. Beginning in 1977, the area of protected forest was expanded in every major watershed. Management plans were made for each protected forest based on their respective classification and characteristics.

At the moment, there are 438,084 ha of protected forests, accounting for 24% of all forest area in Taiwan. The state owns 96% of the protected forests and the private sector only 4%. The system of protected forest is reviewed at intervals of 5-10 years to ensure successful implementation of the system and to provide opportunity for possible adjustments for further improvement (Table 2).

Table 2. **Protected forest area in Taiwan.**

Purpose of protection	Area (ha)	%
Headwater protection	281,178	64.18
Erosion control	128,883	29.42
Sand dune stabilization	6,451	1.47
Landscape beautification	12,353	2.82
Windbreak	3,293	0.75
Flood control	214	0.05
Tide protection	406	0.09
Aquaculture	5,281	1.21
Rock slide protection	25	0.01
Total	438,084	100.00

Source: Forestry Statistics of Taiwan, Taiwan Forestry Bureau, 1991.

3.   Forestry: Past and present

In the preindustrial period about 30-40 years ago, income from natural forest exploitation formed a major part of the government budget. When industrial development gained momentum 20 years ago, the reliance of government on income from natural forest exploitation subsided rapidly. Reservoir siltation in upper mountainous areas become serious, however, which accelerated reafforestation programmes in this period. In the 1970s, pollution and the environmental movement awoke people to concern on the role of the forest in the environment and forest conservation.

Public pressure on issues of nature conservation and environmental protection has resulted in the establishment of more conservation areas and national parks, in which forestry practices are severely restricted.

The forest is a renewable resource. With careful planning and management, forests can provide recreational opportunities, watershed protection, wildlife habitats, timber production, and nature conservation. Unreasonable conservation may result in improper management of forest resources and thus waste a scarce treasure.

4.   New Forestry Management Policies

In 1975, the Executive Yuan (Cabinet) of the ROC announced three principles of forestry management for Taiwan: sustained yield, conservation utilization, and public welfare. Since then, forestry activities followed the Taiwan Forestry Management Reformation act (1976) until 1990. Currently, due to the urging from certain environmental protection groups and the common appreciation of the importance of forests the Taiwan Forest Management Administration Act (TFAMA), 1991, has been approved by the

Executive Yuan and become effective in July 1991. The TFAMA can be summarized into three management categories:

1) The cultivation of plantations: In addition to reafforestation on the cut sites and grassland, emphasis should be placed on timber stand improvement operations and the tending of existing manmade plantations.

2) The protection of forest resources: The administration of protected forests, nature conservation or preservation areas, and the prevention of forest fires and other unauthorized activities within natural forests are the most important tasks. Timber harvesting operations have also been seriously restricted to less than 200,000 cubic metres per year.

3) The implementation of multipurpose forestry management: This includes the development of forest recreational areas, the protection and management of watersheds, and the protection and rehabilitation of forestry research.

The first stage for implementing this new act is six years, from July 1991 to the end of June 1997.

5. National Programme Relating to Forestry Management

Although the forestry policy indicates that forestry management should emphasize resource conservation and forest land protection, this does not mean that all forests should be off limits to forestry operations. Instead, multiple-use management is an ideal way to handle Taiwan's forest resources. The most important forestry activities that are conducted or to be implemented in the following six years are described below.

Cultivation of forests

Establishment and tending of forests are very important because they serve as the basis for creating forest resources. Two methods for cultivating forests are under consideration by foresters. One is to establish artificial plantations, or so-called manmade forests. The other is to improve existing natural regenerated forests. A five-year reafforestation programme (FY 1987 to 1991) has been completed by the Taiwan Forestry Bureau and prefectural governments. A follow-up six-year forest cultivation programme to be initiated from FY 1992 has planned by the Council of Agriculture and approved by the Executive Yuan. The major subprojects of this six-year programme are shown in Table 3.

Table 3. **Subprojects of the forest cultivation programme (FY 1992 to 1997).**

Subproject	Area of quantity	Proposed budget (NT$1,000)*
1. New plantation establishment	16,700 ha	613,297
2. Supplemental planting	8,370 ha	163,858
3. Timber stand improvement	16,690 ha	1,197,702
4. Tending of plantations	378,931 ha	4,143,882
5. Thinning of plantations	6,869 ha	435,900
6. Seedling production	3,066,000 sq m	1,061,185
7. Coastal sand dune stabilization	500 ha	192,100
8. Miscellaneous fees	6 yr	408,986
Total		8,217,000

*US$ 1 = NT$ 26 in 1991.

Source: Implementing Forestry Projects for Water and Soil Conservation Program (FY 1992 to FY 1997), the Council of Agriculture, Executive Yuan, 1992.

Multiple-use management of forest resources

To realize timber and nontimber benefits, forests must be managed properly and related closely to human life. Several functions of Taiwan's forests are especially important, such as forest recreation, slope protection, farmland windbreaks, watershed management, natural ecosystem conservation, environmental greenery, nature education, wildlife management, and timber yield. Timber harvesting during the past decade has declined due to the conservation policy (Table 4). Therefore, nontimber benefits should be emphasized in the future.

Table 4.   **Domestic timber production in Taiwan.**

Year	Sawn timber	Pulpwood	Total
1981	529,138	58,095	587,779
1982	494,937	52,588	547,525
1983	616,070	73,994	690,064
1984	562,637	69,264	631,901
1985	474,584	90,607	565,191
1986	498,674	117,299	615,974
1987	422,644	71,720	494,364
1988	253,312	58,839	312,151
1989	157,289	27,254	184,543
1990	113,830	30,987	144,279

Source: Forestry Statistics of Taiwan, Taiwan Forestry Bureau, 1991.

    A significant project for such nontimber forest resource management is the development of forest recreational area project (FY 1992 to 1997). The objectives of this project are to establish and equip at least 20 forest recreational areas and to provide recreational opportunities for at least 10 million visitor-days annually after the completion of the project. The proposed budget for this project is NT$2,644 million.

Establishment of a forestry geographical information system (GIS)

In order to plan an ideal programme for multipurpose forest management, it is necessary to classify forest lands. To determine whether a specific land class is suitable for timber production, water resource management, wildlife management, nature conservation, or recreational activities, essential information with respect to economic, biotic, and abiotic aspects are required. The forestry geographical information system (GIS) serves as the basis for such land classification, and a land use survey programme was initiated from FY 1993. The GIS is proposed to be completed by FY 1997. At the same time, a draft system for forest land classification and land use planning will also be completed. Total budget for this project is NT$205 million.

6.   Outlook for Forest Resource Management in the ROC

Establishment of intensive forest management areas

Currently, there are some 260,000 ha of manmade plantations within the national forests. Foresters are concentrating their efforts on enhancing land productivity and upgrading the quality of these plantations. On the other hand, private forestry has long been overlooked by the government due to its relatively small proportion of the entire forest area. Owing to public pressure to conserve national resources, timber harvesting in national forests is restricted severely. However, the demand for timber materials is continuously increasing. In order to solve this conflict, strengthening the management of privately owned forests is the preferred method. Therefore, the Council of Agriculture has recently approved a proposal to increase the incentive for private-sector reafforestation from NT$2,000 to NT$32,000 per ha, effective in FY 1991.

All these efforts are aimed at establishing a substantial amount of plantations for intensive management to maximize the production of timber in order to fulfil the increasing demand from society. The primary goal of such intensive management areas is 400,000 ha of plantations.

Implementation of farmland afforestation and agroforestry programme

Overproduction of rice has caused serious problems in Taiwan's agriculture in the past decade. The government has had to provide supplementary fees for farmers to discontinue cultivation in their paddy fields. Meanwhile, the growing demands of society to expand green resources has forced the government to consider alternatives for farmland utilizations. The most recent policy to cope with the above problems is the incentives for farmland afforestation project.

Taiwan is also known as "Formosa," the beautiful island. Accompanying the rapid economic development and the continuous growth of the population, forests have been eliminated from the plains. Furthermore, many slopelands have also been cultivated for agricultural uses, which is potentially threatening to people living in downstream areas. Therefore, the primary purpose in implementing the farmland afforestation project is not only to grow timber for domestic industrial use but also to improve environmental conditions islandwide, especially in the plains.

A pilot incentive farmland afforestation project was initiated in July 1991. The major activities of this pilot project are:

1)   to conduct overall planning for islandwide farmland afforestation;

2)   to demonstrate farmland afforestation operations and agroforestry techniques on some 1,500 ha; and

3)   to raise tree seedlings for FY 1992 and 1993 projects.

The goal of this farmland afforestation project is to create 100,000 ha plantations in abandoned paddy fields, fruit orchards, and other farmlands.

Enhancement of forestry research

It is vital to improve forestry technology in order to modernize forest management, enhance productivity, improve forest products, and protect the environment. Research is essential for technology improvement. To cope with the tasks set by Taiwan's forestry policy, the following research and technology improvement topics are considered as the most obvious and highest priority at this time.

1)   Developing a forest management system for protected forests: The forests especially important for conservation of water, land, and environment are designated as protected forests, and they are under a special controlling system, banning or limiting forest land development. However, regenerating old growth and fostering healthy young stands in protected forests are also important. A selective cutting system may be developed for such regeneration purposes.

2)   Improving the control of forest composition: This topic includes the control of species mix, stand density, and age structure in a given forest. Thinning in manmade plantations and timber stand improvement in natural forests are two major techniques. In addition, mixed forest that contains both hardwood and conifer species is desirable to meet ecological rules. A practical operation system must be set up after a series of field studies to provide the necessary technology for forest managers.

3)   Rationalizing the evaluation of forestry operations: Forest management practices can always disturb the natural situation in forests. Forest managers carry out operations in forests to obtain both timber and nontimber benefits. Therefore, whether a given practice results in a positive contribution to society or not should be evaluated from the economic and noneconomic viewpoints.   Economic evaluation and environmental impact evaluation of forestry operations are two "high-priority" topics.

4)   Management of forest recreational areas: Forest recreation is a major sector of forest management. In order to provide outdoor recreational sites without damaging resources, several problems must be solved. First, visitor preferences for forest recreational areas, particularly high-use sites accessible to the urban population must be identified. Second, managers should determine the social, economic, and environmental uses and developments. Third, techniques for managing recreational sites at an acceptable carrying capacity level must be evaluated in advance.

5)   Improving forest land productivity: The world demand for forest products in the coming decades is projected to increase continuously while the land area available to forestry will decline during the same period. Thus it is necessary for wood importing countries such as the ROC to improve the productivity of their own forest land to increase timber inventory for wood-related industry in the future. Measures under consideration for improving forest productivity are genetic manipulation, thinning, timber stand improvement, etc.

7. Conclusions

The forest is a natural resource and part of the human environment. Presently Taiwan is at a turning point in its forest management. The welfare of the people, the quality of life, and the conservation of the environment are important issues being faced. Therefore, the forest management programme is no longer tied to timber harvest receipts alone. To achieve successful forest management, technical support from research and development (including extension) activities are essential. Initiating an integrated research programme to coordinate all research agencies and researchers is a basic step toward the success of forest resource management. Also, experienced forest managers and researchers should work together to create applicable plans for resource conservation and operations to improve the effectiveness of forestry resource management.

COUNTRY PAPER: HONG KONG (1)

Man-Kwong Cheung
Senior Conservation Officer
Agriculture and Fisheries Department
Hong Kong Government
Hong Kong

1.   Introduction

Although Hong Kong is regarded as one of the world's busiest cities, about three-quarters of its 1,075 square kilometres is countryside. The overall objective of nature conservation in Hong Kong is to conserve the countryside by the effective protection and management of its landscape, vegetation and wildlife for the benefit of present and future generations, and to control the trade in endangered species of animals and plants in accordance with the provisions of the Convention on International Trade in Endangered Species of Wild Fauna and Flora (CITES).

As a department responsible for forestry and countryside conservation, the Agriculture and Fisheries Department carries out various activities to protect and manage important countryside habitats, in particular woodland and vegetation covering hillsides. Continual surveillance is maintained on local plant and wildlife habitats, especially those designated as sites of special scientific interest, against disturbance. The trade in endangered species of fauna and flora is regulated by licensing and law enforcement.

One important achievement in nature conservation in Hong Kong is the establishment and management of the country parks estates, which covers about 40% of the total land area of Hong Kong. The country parks are managed on a multiple-use basis for water conservation, nature conservation, and countryside recreation. Country park staff are deployed to manage woodlands and to protect the vegetation against fire, to control development within country parks, to keep country parks clean and tidy, and to promote better understanding of the countryside and its value as a resource for recreational, educational, and scientific purposes.

Another important development over the last decade has been a greater awareness among the general public of the importance of nature conservation and the need to conserve flora, fauna, and their natural habitats. Countryside education and interpretation have become important and integral parts of the conservation and country parks programme. The long-term support and success of nature conservation will depend to a certain extent on the efforts made in this respect.

The role of forestry in the overall economy is relatively insignificant and its contribution to the community is in the form of soil and water conservation, countryside recreation, nature education, and amenities rather than timber and fuel production. The felling of tress for timber or fuel is not significant in Hong Kong. Instead, more tress are planted in the countryside, country parks, and urban fringe areas. Roadside planting and landscape planting in housing estates have also become very popular despite the rather congested city area in the territory.

This paper briefly describes activities in the above areas.

2.   Countryside Conservation and Forestry in Hong Kong

Organization and administration

The Agriculture and Fisheries Department is the forestry and countryside conservation authority in the territory. Within the department, the Conservation and Country Parks Branch, headed by an assistant director, is specifically responsible for forestry, nature conservation, and related activities. The branch consists of three divisions, the Conservation Division, the Country Parks Division, and the Countryside Development Division. The Conservation Division is responsible, inter alia, for the general conservation of flora and fauna including the control of international trade in endangered species and the production of tree stock for afforestation. The Country Parks Division is responsible for the establishment, management, and protection of forest plantations and woodlands as well as the provision of informal outdoor recreational facilities in the 41,320 ha of country parks. The Countryside Development Division, in addition to dealing with planning and recreational development work in country parks, is also concerned with the management of woodlands in the open countryside not designated as country parks. The three divisions work closely together to implement policies relating to the conservation of flora and fauna and the planning, development, protection, and management of forestry areas, country parks, and nature reserves.

Apart from the Agriculture and Fisheries Department, there are six regional landscape core teams, each consisting of landscape architects and forestry officers together with supporting staff in the Territory Development Department. Among other landscaping work in the new towns, these teams are responsible for the initial establishment of woodlands outside country parks for the purposes of soil and water conservation and landscape rehabilitation. Amenity woodlands are eventually handed over to the Agriculture and Fisheries Department for management.

3.   Nature conservation and Forestry

The overall countryside conservation policy is to protect, conserve, and extend natural woodland and forest plantations for multiple uses including environmental improvement, countryside recreation, soil and water conservation, landscape rehabilitation, and outdoor education. The forests are not managed for timber or firewood production. As the actual monetary return from forests is not important, Hong Kong can afford to place more emphasis on the community and social output of its existing forests. This policy has been adopted since the early 1970s when the need to conserve the countryside was widely recognized. In the following years a vigorous country parks programme was launched for the purposes of countryside conservation, recreation and education. By now 21 country parks together with 14 special areas have been designated under the Country Parks Ordinance. These occupy a total of 41,320 ha of land, which is about 40% of the total land area in Hong Kong (Appendix 1). In past years afforestation and landscape planting have become an important part of the government's policy for rural planning and development projects. More emphasis has now been placed upon landscape restoration for borrowing, controlled tipping, roads, building works, and similar engineering projects. Urban forestry, roadside planting, and preservation of existing vegetation cover including urban trees have become integral parts of forestry in Hong Kong.

4.   Forestry Resources and Financing for Forestry and Nature Conservation

In Hong Kong, forests are mainly managed for the social needs of the people rather than for tangible material returns. The value of forestry lies in its contribution to recreation, nature conservation, environmental restoration, and outdoor education. Thus forests are now regarded as an important social and environmental asset in terms of government policy. There is no primary forestry industry in Hong Kong. Apart from a small amount of thinning that has been used for fencing and recreational facilities in country parks, there are no commercial forest products. All timber consumed in Hong Kong is imported, and semifinished or finished timber products are exported.

All funds for forestry, country parks, and conservation activities come from the government through:

1)   the conservation and country parks programme;

2)   the new town development programme; and

3)   engineering projects.

Until the present, there has been no major problem in funding the establishment of new plantations, although it has been rather difficult to obtain funds for recurrent maintenance. Apart from very small proceeds from the sale of young tree seedlings, there is no direct revenue from forest plantations. In this regard Hong Kong is self-sufficient and there has been no external financial aid.

5.   Afforestation and Forest Management

Extension of plantation areas

In the past 10 years a total of about 4 million seedlings of various species have been planted for amenity, soil conservation, landscape rehabilitation, and for replacement of fire damage within country parks. Over 80% of the trees and shrubs planted were broadleaved species, and mixed planting was used whenever possible. A list of major afforestation species is given in Appendix 2. Silvicultural treatment including pruning, thinning, clearing of undergrowth, and fertilizing have been carried out in the plantations to improve forest hygiene and to promote tree growth. Most of these woodlands are located on government land and they are established for erosion control, amenity or for landscape rehabilitation of engineering projects.

Protection against fire

Fire is undoubtedly the most severe problem for the forestry authorities. In the past 10 years the average annual fire incidence has been in excess of 2,600 outbreaks, with about 60% inside or threatening country parks or plantations. The worst fire occurred in the Shing Mun and Tai Mo Shan country parks on January 8-9, 1986, damaging 282,500 trees and 750 ha of hill land. This was the most serious fire since 1979. The fire problem significantly influences the organization and scale of provision of management services. It also influences the road construction programme, the selection of species for afforestation, and the siting, type, and scale of recreational facilities to be provided.

The interdepartmental Working Group on the Fire Prevention Publicity Campaign has continued to arrange publicity on preventing countryside fires through the mass media, the distribution of printed publicity materials, and mobile displays organized at different venues. Visits to local communities are made regularly by park wardens to remind villagers to be careful when burning weeds or rubbish in their fields. The intensive fire prevention publicity over the years and the provision of a large number of barbecue pits for picnickers has resulted in greater public awareness of the damaging effects of hill fires. Those responsible for the illegal lighting of fires in the countryside are prosecuted. In addition, fire prevention operations, including the establishment of firebreaks by means of grass cutting and controlled burning, and the reduction of fire hazards by thinning and pruning, have been continued in forest plantations. More water tanks have been constructed in plantations to facilitate firefighting.

Pests and diseases

The infestations of pine wilt nematodes on the South China pine (_Pinus massoniana_) is spreading to most parts of the territory. The cut-and-burn approach was adopted to contain the problem. These pines were planted in the early 1950s and 1960s as "pioneers" on exposed, infertile hill land to create a more favourable environment for the development of broadleaved tress and shrubs. The mixed broadleaved vegetation is more resistant to fire and pests than pines, and was an advanced form of vegetation in the process of natural plant succession in Hong Kong. The gradual dying off of the pines has accelerated the succession process. As pinewood has no market value in Hong Kong, the loss of pine plantations does not cause financial hardship to their owners.

Tree seedling production

During the past 10 years, a total number of 7.6 million seedlings have been produced in tree nurseries for various government afforestation and landscaping programmes. More recently the average production has been about 0.8 million per year, among which half of the seedlings have been issued for afforestation work in country parks while the remaining have been supplied to other government departments for landscaping projects in new towns, along highways, etc. To encourage the planting of trees and shrubs by the general public, free seedlings are issued to schools and charitable organizations for educational and environmental improvement purposes while a small portion is sold to the public for general amenity planting in private areas.

Developments in related fields

In order to study the adaptability of local and exotic trees and shrubs under different site conditions, a series of trials has been carried out in past years to compare their growth rates under various conditions. One of the findings is that the two exotic species of *Acacia*, i.e., *A. auriculaeformis* and *A. mangium* adapt very well under local conditions. They can serve as pioneer species and provide fast vegetation cover to disturbed and badly eroded soil.

The first local arboretum, in the Shing Mun Country Park, with a collection of over 300 species has been open to visits by schoolchildren since 1985. The plant collection there helps students and botanists as well as local naturalists to gain a better understanding of the tree species in the territory. A second arboretum is being developed in Sai Kung together with a specimen orchard with a rich collection of local fruit trees. These arboreta are open to the public, especially to school parties, for the purposes of outdoor education and for academic studies.

6. Establishment and Management of Country Parks

To conserve and, where appropriate, open up the countryside for the greater enjoyment of the population, the Country Parks Ordinance was enacted in 1976 to provide a legal framework for the designation, development, and management of country parks and special areas. It provides for the establishment of a Country Parks Board to advise the Director of Agriculture and Fisheries who, as country parks authority, is responsible for all matters on country parks and special areas. Country parks are designated for the purposes of nature conservation, informal recreation, and outdoor education. Special areas are created mainly for the purpose of nature conservation.

A total of 21 country parks and 14 special areas have been established, covering a total area of 41,320 ha. The country parks comprise scenic hill lands, woodlands, reservoirs, islands, and coastline in all parts of Hong Kong. The parks are now very popular with all sections of the community and a day in the countryside is an accepted part of the recreational opportunities available to the people. Over 9 million visitors were recorded in 1991 and activities ranged through leisure walking, fitness exercises, hiking, barbecuing and family picnics, and camping. Facilities provided in the parks include picnic sites with tables and benches, barbecue pits, litter bins, children's play apparatus, campsites, and toilets, all carefully designed to blend in with the natural environment.

Footpaths, family walks, and nature trails provide easy access to the hills and the woodlands to allow visitors to take every opportunity to enjoy the scenic beauty of these areas. Guidebooks for nature trails are available, and major paths are being improved and waymarked through the hilly terrain. Increasing emphasis is being given to facilities to help visitors to enjoy and understand the countryside. In this connection, six visitor centres have been established in Aberdeen, Tai Mei Tuk, Pak Tam Chung, Tai Hang Tun, Shing Mun, and Tai Mo Shan. The Tsiu Hang special areas located near Pak Sha Wan in Sai Kung, which consists of a rich collection of fruit-bearing and amenity trees, have been developed into a nature education centre for the purpose of nature education.

As mentioned above, fire is the major hazard and it bedevils park management for about six months every year during the cool, dry winter when many people like to spend a day out, especially weekends and on public holidays, to walk in the hills. Litter is another problem. It is one of the tasks of park management to provide for the collection of such litter from the countryside, which in 1990 totalled some 4,000 tonnes. With such problems in mind, the Country Parks Authority has provided barbecue pits and litter bins located strategically throughout park areas for the use of visitors. The authority also prosecutes anyone found littering, damaging facilities, or lighting fires outside the approved barbecue sites in country parks.

7. Countryside Education and Community Involvement

The Agriculture and Fisheries Department attaches great importance to countryside education. Over the years, it has provided facilities and organized activities for countryside education and for physical involvement in countryside management work for members of the public. It is believed that through active participation in and better understanding of nature, the general public will develop a sense of respect and love for the countryside and an improved awareness of the need of its conservation. To cater for thousands of visitors flocking to country parks each weekend, six country park visitor centres and nature education centres have been established in strategic locations to provide them with regular information and other educational materials on the countryside. Regular guided walks are organized at various country park areas for schools and organized groups. Self-guided nature trails with guidebooks are established in various parts of the country park system.

In 1991, about 110,0000 people participated in slide shows, lectures, educational walks, litter collecting, tree planting, and various forms of voluntary conservation work organized by the Agriculture and Fisheries Department or in conjunction with other departments or organizations.

Various aspects of advice and assistance were offered to the general public, including country park visitors, schools, and students on special projects. Annual educational activities organized by the department include the community tree planting scheme that attracts some 20,000 participants annually, over 100 forestry work camps to enable 2,000 schoolchildren and youths to learn and practise simple countryside management work during their summer holidays, and training camps for student conservation leaders. The "Clean and Green Scheme" is both fun and educational and has remained very popular since its launching in 1982. Schools and organized groups can join the "forest adoption scheme." Each school/group is assigned a small plot of land within country parks on which students carry out tree planting, simple forestry work, and field studies.

In addition, there is growing public concern and support for environmental conservation. Local conservationists, and academic and conservation groups have occasionally offered valuable advice and proposals to the government. Many have joined forces with the government in various aspects of countryside conservation work, including participation in tree planting activities and organization of educational programmes. Other service clubs, such as the Lions Club and Rotary Club, are enthusiastic in supporting countryside conservation and have sponsored a variety of activities in country parks in recent years.

The management of Mai Po marshes is a good example to illustrate the partnership between the government and nongovernmental organizations in achieving countryside conservation objectives. The World Wide Fund for Nature/Hong Kong (WWF/HK) is developing and managing the Mai Po marshes for conservation and educational purposes. To manage the area better, shrimp ponds (*gei-wais*) are gradually being acquired by the WWF/HK from public donations and fund-raising activities. It has set up a wildlife education centre and a study centre, and regularly organizes guided walks for schoolchildren and members of the public. The government is responsible for maintaining restricted access to the nature reserve, taking enforcement actions to curb abuses of conservation legislation, and controlling incompatible development in or around the areas.

8. Conservation Legislation and Enforcement

Four pieces of legislation have been enacted specifically for the conservation of flora, fauna, and natural habitats. They are briefly described below.

1) Forests and Countryside Ordinance Cap. 96 and subsidiary legislation: The ordinance and its regulations protect forests, trees, and important plants from damage, and collection, and the illegal use of fire in open countryside is also prohibited.

2) Wild Animals Protection Ordinance Cap. 170: The ordinance gives full protection to all local wildlife by prohibiting hunting territorywide. Possession, export, and sale of protected local wild animals are prohibited under the ordinance.

3) Country Parks Ordinance and its subsidiary Regulation Cap. 2087: The ordinance and its regulations protect flora, fauna, and habitats from damage, disturbance, or hunting. They also control development and activities within country parks.

4) Animals and Plants (Protection of Endangered Species) Ordinance Cap. 187: The ordinance restricts the import, export, and possession of endangered species and their readily recognizable parts and derivatives in accordance with CITES.

Regular patrols are conducted by country park rangers and nature wardens throughout the country parks system to enforce the above ordinances.

9. Acknowledgments

The author wishes to express his gratitude to the Asian Productivity Organization for providing the opportunity and the Director of Agriculture and Fisheries of Hong Kong for his kind permission to attend this Study Meeting on New Trends in Environmental Management in Japan. The opinions and views expressed in this paper are those of the author, and they may not represent the official points of view of the department.

Appendix 1. Country parks and special designated areas in Hong Kong.

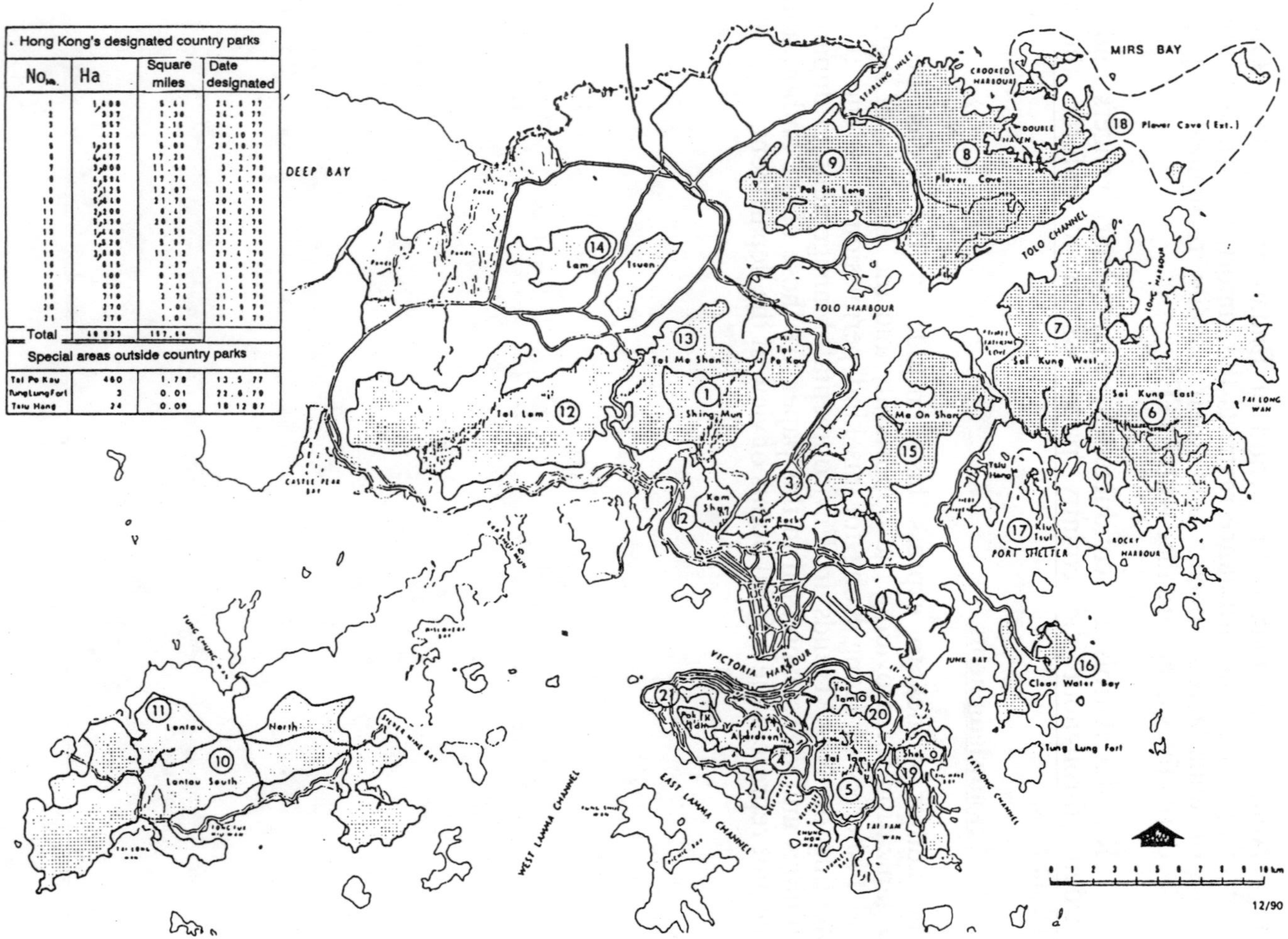

Hong Kong's designated country parks			
No.	Ha	Square miles	Date designated
1	1,400	5.41	24.6 77
2	337	1.30	24.6 77
3	557	2.15	24.6 77
4	423	1.63	28.10 77
5	1,315	5.08	28.10.77
6	4,477	17.29	3.2.78
7	3,000	11.58	3.2.78
8	4,594	17.74	7.4.78
9	3,125	12.07	18.8.78
10	5,640	21.78	20.4 78
11	2,200	8.49	18.8.78
12	5,330	20.58	23.2.79
13	1,440	5.56	23.2 79
14	1,520	5.87	23.2.79
15	2,880	11.12	27.4.79
16	615	2.37	28.9.79
17	100	0.39	1.6 79
18	630	2.43	1.6 79
19	710	2 74	21.9 79
20	270	1.04	21.9 79
21	270	1.04	21.9 79
Total	40 833	157.66	
Special areas outside country parks			
Tai Po Kau	460	1.78	13.5 77
Tung Lung Fort	3	0.01	22.6.79
Tsiu Hang	24	0.09	18 12 87

Appendix 2.   Major afforestation species in Hong Kong.

Acacia auriculaeformis

Acacia confusa

Acacia mangium

Castanopsis fissa

Casuarina equisetifolia

Cinnamomum camphora

Eucalyptus torelliana

Gordonia axillaris

Liquidambar formosana

Machilus breviflora

Melaleuca leucadendron

Pinus elliottii

Sapium discolor

Schima superba

Tristania conferta

COUNTRY PAPER: HONG KONG (2)

Bernard Ian Dubin
Senior Environmental
 Protection Officer
Environmental Protection
 Department
Hong Kong Government
Hong Kong

1.   Introduction

This country paper presents information on the situation in Hong Kong with regard to the following issues: the use of tropical rainforest products in Hong Kong and some recommendations to try and reduce deforestation in the region; current recycling initiatives including progress on the recycling of waste paper and construction waste; and recent progress in introducing the concept of environmental management in Hong Kong.

2.   Forestry resources and Hong Kong

In classical economic terms, that is, in terms of simple monetary value, there are no native forestry resources in Hong Kong. While there are wooded areas in the Territories, these resources are exploited strictly as educational conservational, and recreational resources. Of course, from the viewpoint of environmental economics, these are very important forestry resources. Hong Kong's country parks cover some 40% of the Territories area and contain most of the wooded areas in the Territory. These areas are protected from development by the Country Parks Ordinance.

Tropical rainforests

Worldwide, the depletion of the tropical rainforests has been recognized as a problem of global import. Tropical rainforests are probably of great significance to global climate patterns and are a storehouse of biological diversity. Newsweek for December 31, 1990, quoted a US EPA Science Advisory Board report that identified four issues of "high risk environmental priority" for the decade:

1)   climate change;

2)   ozone depletion;

3)   destruction and alteration of wildlife habitats; and

4)   species extinction.

All four of these global issues can be related to tropical forest depletion. The issues include the rainforest link to the release of cloud-forming moisture and the connection with planetary systems regulating the amount of carbon dioxide in the atmosphere and affecting the greenhouse effect. Some global warming scenarios predict havoc on a grand scale as ice caps melt and low-lying regions flood. There is very little argument with the thesis that preservation of the rainforests is in the best interests of mankind as a whole.

The Hong Kong link

According to information obtained by Friends of the Earth (FOE) from the International Tropical Timber Organization, Hong Kong is one of the world's major importers of tropical timber. In 1990, Hong Kong ranked fifth behind France, South Korea, China, and Japan for imports of rainforest logs (FOE, 1992). Table 1 shows recent figures on Hong Kong lumber industries imports and retained imports for the last three years for which figures are available. FOE has determined that 1988 imports of tropical hardwood were sufficient to cover a block the size of a football field to a depth of 174 metres. East Malaysia and Indonesia are the primary sources, although there are significant imports from other countries including Burma, Thailand, the Philippines, Vietnam, and China. Some of these imports have been associated anecdotally with human rights violations in Burma.

Table 1. Information on Hong Kong lumber industry imports and retained imports of wood.

	1988 Imports	1988 Retained imports	1989 Imports	1989 Retained imports	1990 Imports	1990 Retained imports
Teakwood, in the rough or roughly squared (M_3)	18,634	9,648	26,915	15,626	20,842	10,524
Sandalwood, in the rough or roughly squared (kg)	440,467	3,219	259,512	89,886	322,577	45,225
Decorative woods, n.e.s. (blackwood and rosewood), in the rough or roughly squared (m3)	7,603	3,542	14,391	7,787	24,586	15,709
Sawnlogs and veneer logs, of nonconiferous species, in the rough or roughly squared, n.e.s. (m3)	593,351	547,653	517,140	420,854	526,566	458,282
Teakwood (conversions and squares) (m3)	17,074	7,433	22,284	8,676	20,609	12,143
Sandalwood, simply worked (kg)	25,961	23,058	41,980	26,823	6,100	-2,336
Lumber of nonconiferous species, sawn, planed, grooved tongued, etc., n.e.s. (m3)	236,032	73,167	222,269	26,422	206,802	19,091

(Industry Department figures, January 1992)

The building industry in Hong Kong has traditionally used hardwoods (because they were the indigenous material) for formwork "shuttering" structures for moulding reinforced concrete. This timber formwork is reused several times until it has lost its ability to be reworked economically and is then either burned for rough fuel purposes or disposed of at a landfill as construction waste. In 1988, 300 tonnes of timber per day were discarded in this fashion. It is recognized that softwoods are just as effective for formwork purposes and a good deal cheaper, although environmentalists have failed to get this point over to the industry.

Conservation groups in Hong Kong have recommended that a contribution to reducing this demand could be made and at the same time reduce costs in the building industry. While the market mechanism has not worked so far, it may be that through a simple expedient such as inserting a clause in all government building contracts to the effect that soft, not hardwoods should be used for formwork, it could be possible to trigger an increasing demand for softwood. The government is considering introducing policies in line with these views.

A recent FOE report recommended that Hong Kong's construction industry and government departments should:

1) formally adopt a stated policy requirement that reusable steel formwork be employed in future projects;

2) formally adopt a stated policy that where one-off casts are required, tropical woods should not be used and softwood alternatives be used instead;

3) formally adopt a stated policy preference for contractors employing reusable steel formwork instead of wood, and, where wood must be used, employing softwoods;

4) amend specification requirements of hardwoods for interior use in favour of softwoods or other materials; and

5) contractually require contractors not to use wooden formwork for shuttering except for one-off irregular shapes.

In a related matter they have also recommended Government introduce charges for construction waste dumped at landfills.

3. Waste recycling in Hong Kong

In 1990, municipal solid waste disposed of in Hong Kong amounted to 16,000 tonnes per day (tpd) (Fig. 1) (Cheung and Rootham, 1991). The main components were domestic waste (approximately 5,400 tpd), industrial waste (approximately 1,300 tpd), commercial waste (approximately 400 tpd), construction waste (approximately 8,400 tpd), and special waste (approximately 400 tpd). These wastes were disposed of either by incineration plants or at landfills. The quantity of solid waste requiring disposal has been increasing at a rate of 10% per annum over the last 16 years, roughly in line with gross domestic product (GDP) growth (Fig. 2). As a result of the recent construction boom, construction waste has increased significantly. The average quantity of construction waste disposed of at landfills increased from 5,600 tpd in 1989 to 8,600 tpd in 1990. The Environmental Protection Department commissioned a consultancy study in March 1991 to develop appropriate construction waste recycling strategies. The final report of this study has recently been accepted and is discussed below.

Current recycling situation

Present recycling initiatives are entirely dependent on private enterprise, from collection through separation and reprocessing or export of waste components to marketing and sale of semifinished or recycled materials. Existing waste recycling activities are export oriented and play a significant role in waste management. In 1990 (Cheung and Rootham, 1991) a total of 1.22 million tonnes of waste was exported, amounting to 17% of the total solid waste generated in Hong Kong. Waste paper and alloy steel are the two major components (averaged 1,500 tpd and 1,200 tpd, respectively); the others are nonferrous metals, plastics, and some textile fibres (Fig. 3). The export of waste accounted for an earning of $2.3 billion. Apart from these earnings, these recycling activities contribute directly toward reducing the demand for the limited landfill space available in Hong Kong. Other local recycling industries include plastic scrap recyclers, tyre retreaders, and precious metal refineries.

While waste reprocessing operations in Hong Kong are predominantly small scale, several large-scale recycling operations have commenced within the last few years. These include two waste paper recycling plants and a used lubricating oil recovery factory that have opened since early 1990. Approximately 430,000 tonnes of waste paper, used lubricating oil, metals, plastic scrap, and glass were reprocessed locally in 1990 (Cheung and Rootham, 1991).

Waste paper recycling initiatives

Recently the government has been actively promoting the separate collection of paper waste, chosen because of the existence a well-established market as well as the availability of local recycling plants. Paper waste is also easily identified and recovered from the waste stream. A copy of a guide on how to start a waste paper recycling programme in offices is attached and similar guides for residential buildings have been published.

To date, over 60 government departments and branches have initiated their own schemes to collect paper waste. A review of the results of these trials (Cheung and Rootham, 1991) have indicated that the following are the key factors ensuring a high capture rate of waste paper:

1) sufficient number of collection boxes to ensure convenience of disposal by waste producers and collection of scrap paper by cleaners;

2) detailed guidelines on the types of paper that should be separately collected for recycling; and

3) designated persons to attend common areas and be responsible for reviving and sustaining staff enthusiasm in participating in the collection scheme.

In addition, the government in collaboration with the private sector Environmental Campaign Committee (ECC) is launching waste paper recycling schemes in all its housing estates. The private sector has shown a positive response to the government's publicity campaign on paper waste recycling. Over 450 establishments including schools, commercial offices, banks, and social centres have initiated separate collection of waste paper.

To complement waste paper recycling, the government is also actively promoting waste minimization, specifically targeting paper and plastic bags. In addition, the use of recycled paper stationery is being examined with a view to establishing government procurement policy and guidelines for recycled products.

Construction waste recycling

Hong Kong's landfills are filling up at a frightening rate and a means must be found to reduce the amount of material arriving at them for disposal. The Study on Recycling of Construction Waste Received at Landfills was carried out by a project team comprising Donahue/JRP Asia Pacific Ltd (Donahue/JRP) in association with Gershman, Brickner & Bratton, Inc. (GBB). The objectives of the Study were as follows:

1) to evaluate overseas experience in construction waste recycling with respect to Hong Kong conditions;

2) to identify the nature of potentially recyclable products from construction waste received at landfills and to assess the availability of markets/outlets for these products;

3) to verify the practicality of processes by pilot testing or other suitable means;

4) to study the institutional arrangements for the application of potential recycled products at landfills; and

5) to recommend a system for recycling construction waste at a landfill in Hong Kong giving a full account of the economic, engineering, operational, and environmental considerations.

Table 2. **Potential recovery quantities of construction waste constituents.**

Constituent	Recovery quantity in tonnes		
	Concrete and rock plant (Option 1)	Mixes construction Waste plant (Option 2)	Combined system (Option 3)
Asphalt	33,398	8,997	42,395
Concrete	222,173	59,849	282,022
Reinforced concrete	246,013	66,271	312,284
Dirt	358,878	193,348	552,226
Rock	173,629	46,772	220,401
Rubble	116,166	31,292	147,458
Wood	0	0	0
Bamboo	0	0	0
Block concrete	12,101	3,260	15,361
Brick	78,057	42,053	120,110
Glass	4,808	2,590	7,398
Other organics	0	0	0
Plastic pipe	0	0	0
Sand	47,757	25,729	73,486
Trees	0	2,289	2,289
Fixtures	0	0	0
Junk	0	0	0
Metal (ferrous)	21,758	51,480	73,238
Residue	77,562	270,594	348,156
Total	1,392,300	928,200	2,320,500

(After Donahue/JRP and GBB)

In 1991, the estimated rate of generation of construction waste was 10,675 tonnes per day, an enormous increase from the 1990 figure of under 6,000 tonnes. A field study assessed the constituents of the construction waste received at one of the Territories largest operational Landfills at Tseung Kwan O. It was found that many of the arriving loads are over 90% inert material and there could be potential for profitable recycling operations. Nineteen of the constituents found in construction waste were examined for their potential for reuse (Table 2). Recoverable constituents were identified as follows: asphalt; aggregates; soft fill; ferrous metal; and wood. The study recommends that the five recoverable constituents can be aggregated into three basic groups of recoverable materials which comprise 97% of the construction waste stream. These groups are: inert granular material (aggregate, asphalt, and soft fill); wood; and ferrous metal. The study has determined that if only 85% of the recoverable portion of construction waste is recovered, the equivalent of 2½ months of landfill life can be saved every year. The final report goes on to recommend implementation of a system of construction waste recycling, together with institution of a user fee at all waste disposal facilities. This is in line with recent recommendations made by the FOE.

A pilot scheme will be instituted in March 1992, at Tseung Kwan O landfill to verify the applicability and effectiveness of construction waste recycling in Hong Kong prior to full-scale implementation. In view of the current critical shortage of landfill space, the government has developed a plan to ban all construction wastes at landfills and to require the construction industry to carry out separation/recycling before disposing of material suitable for use as fill at public dumps and unsuitable material at landfills. This proposal is intended to force the development of recycling schemes within the construction industry itself and reduce the demand for the limited and valuable landfill space.

Environmental Management Practices

International trade

Many Hong Kong manufacturers are beginning to encounter environmental clauses in their overseas trade contracts. For example, manufacturers of high-technology products have found that European and North American customers insist on a declaration that chlorofluorocarbons (CFC s) have not been used in their manufacturing processes. The Hong Kong government is attempting to turn such changing trading conditions to Hong Kong's advantage by promoting environmental management techniques within industry. The feeling in the government is that if Hong Kong can stay ahead of the game on environmental issues, it should be able to retain a competitive edge over Asian rivals in the major European and North American markets for high-technology products.

One such initiative has recently come to fruition. The San Miguel Corporation in association with the Environmental Protection Department undertook to fund a consultancy study that has resulted in publication of a manual on environmental audit. The manual will assist companies developing environmental audit programmes aimed at internally reviewing their own processes, procedures, and practices and improving their impact on the environment. Hong Kong companies are realizing that protection of the environment is good business and that pollution prevention pays. The audit manual is directed at assisting companies instituting environmental audit programmes to:

1)   make their operations more efficient;

2)   use materials and to make products that have minimal environmental impact;

3)   reduce waste;

4)   identify potential problems and liabilities before they become critical;

5)   ensure that they always comply with regulations; and

6)   provide a basis for enhanced public awareness of their operations.

Environmental audit is an in-house procedure and it is government policy that the results are an internal company matter. Audit programmes are normally tailored specifically to the needs of individual companies, because they are planned and executed by senior staff within the companies themselves. Outside assistance is not necessary to establish an effective audit programme; in most cases the necessary technical expertise already exists within the company structure. In order to establish an audit programme, the key first step is to identify a member of senior management to become responsible for the development of a company environmental policy. The environmental audit manual, specifically designed for companies in Hong Kong, explains to that person how to integrate environmental management principles as a vital element of corporate business strategy, for the benefit of management, staff, the company shareholders, consumers, and the environment. It provides a clear explanation of environmental auditing, its role in management, and the many benefits that can be achieved through the application of the principles.

Since the manual was released on November 18, 1991, hundreds of copies have been distributed to the largest companies and businesses in Hong Kong. It is understood that many of these 'Hongs' have instituted the necessary programmes and it is hoped that the results will begin to show soon, in tangible reductions in waste and pollution. Corporate sponsorship is currently being sought to undertake production of smaller, simpler guides for specific industries, and in particular for the modest industrial operations that make up the bulk of Hong Kong manufacturing.

Acknowledgements

Most of the information in this paper was obtained from previously published sources. In particular, a great deal of the information on recycling was derived directly from the two references quoted. The paper has not been reviewed within the Hong Kong government, and opinions expressed herein are those of the author alone and do not represent official policy of the Hong Kong government, except as otherwise stated.

References

Cheung, B. M. H. and Rootham, R. C. *Waste Recycling in Hong Kong.* Proceedings of Pollution in a Metropolitan Environment (Polmet). Hong Kong: Hong Kong Institution of Engineers; 1991.

Cheung, B. M. H. and Rootham, R. C. *Report on the Use and Wastage of Tropical Timber by Hong Kong's Construction Industry.* Hong Kong: Friends of the Earth; 1992.

Donahue and GBB. *Study on Recycling of Construction Waste Received at Landfills*, Final Report. Hong Kong: Hong Kong Government Environmental Protection Department; 1991.

Arthur, D. Little Asia Pacific Limited. *Environmental Audit Manual, A Guide to Help Companies Review and Improve Their Environmental Performance.* Hong Kong: San Miguel Brewery Limited; 1991.

Fig. 1. <u>Municipal solid waste in 1990 (16,000 tpd received at disposal facilities).</u>

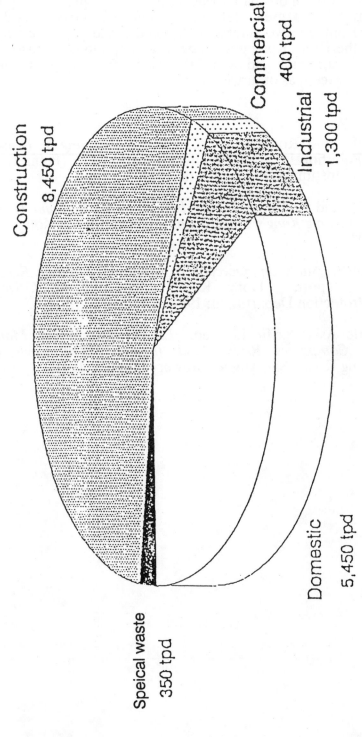

Construction
8,450 tpd

Commercial
400 tpd

Industrial
1,300 tpd

(After Cheung and Roothan)

Speical waste
350 tpd

Domestic
5,450 tpd

Special waste includes waste from abatloirs, hospitals, water/wastewater treatment works, livestock, condemned goods, and chemical wastes.

Fig. 2. Municipal solid waste (MSW) increase with GDP growth.

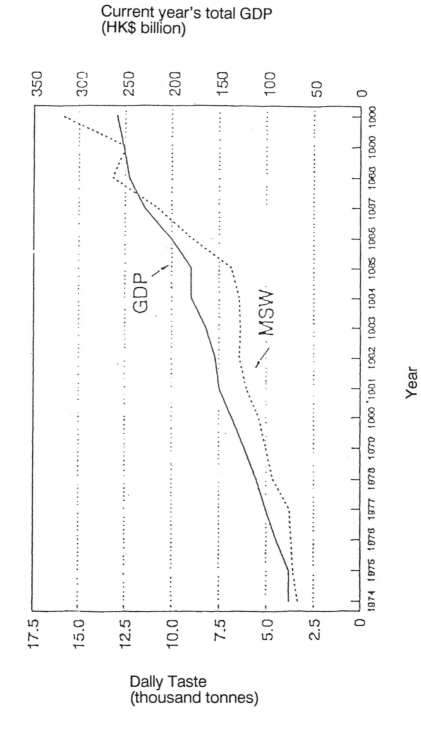

(After Cheung and Rootham)

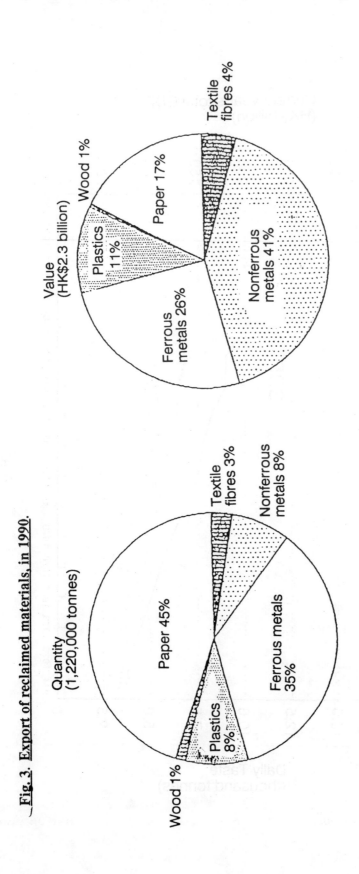

Fig. 3. Export of reclaimed materials, in 1990.

Quantity
(1,220,000 tonnes)

Paper 45%

Ferrous metals
35%

Nonferrous
metals 8%

Textile
fibres 3%

Plastics
8%

Wood 1%

Value
(HK$2.3 billion)

Nonferrous
metals 41%

Ferrous
metals 26%

Plastics
11%

Paper 17%

Wood 1%

Textile
fibres 4%

(After Cheung and Rootham)

COUNTRY PAPER: INDIA

R. V. Krishnan
Principal Secretary
Department of Energy,
Forests, Environment,
  Science & Technology
Hyderabad, India

1.   Introduction

The World Bank has identified five key problems in environmental conservation and ecology worldwide:

1)   destruction of natural habitats;

2)   land degradation;

3)   depletion of water resources;

4)   degradation of global commons; and

5)   urban, industrial, and agricultural pollution.

The first four problems are closely connected with natural resources, particularly forests; hence the following paper focuses mainly on India's forest estate and how well it has fared through the history of the country and its present status.

> "I wonder sometimes if there is any other natural resource which gives us so much and of which we know so little as forest."
>
> Dr. Rajendra Prasad
> (1st President of Independent India)

Forest area and forest diversity in India

India's forests cover (according to latest figures) 64.2 million ha, constituting 19.5% of country's land area. The forest vegetation varies from tropical evergreen forests in the Andaman and Nicobar islands to dry alpine forests high in the Himalayas. Between these two extremes the country has semievergreen forests, deciduous, littoral, swamp, thorn, subtropical broadleaved hill and pine forests, and montane temperate forests.

Species diversity of forests

Species richness is characteristic of all tropical forests. Although tropical forests cover only 7% of the earth's land surface, they carry 50% of all species. On an average a tropical forest contains 216 species per ha against 25 species per ha in a temperate forest. In India 45,000 species of plants occur; several thousand species are endemic to India. A special feature of India is the occurrence of over 113 species of bamboo (considered a sacred plant by Japanese). With regard to fauna, the diversity is impressive with over 350 species of mammals, 120 species of birds, and over 20,000 species of insects.

Variety of uses of forests

Rich as India's forests are in species diversity, so also is the wide variety of uses to which forest products are put. The forests supply not only timber, fuel wood, and bamboo, but also fruits, fibre, gum, honey, tussar silk, brooms, etc. and a tremendous variety of medicinal herbs used as raw material for several Western medicines as well as in Ayurveda (the ancient Indian system of medicine and science of long life). The list is only illustrative as there are myriads of uses to which forest products are put.

2.   Brief Survey of Forest Conservation in India

Forest conservation in India as elsewhere in the tropics is marked by a wide gap between aspirations and efforts, between preaching and practice. The Agni Purana enjoined on its devotees 4,000 years ago to earn religious merit and material prosperity by protecting trees. The Gautama Buddha 2,500 years ago preached that every person should plant one tree every five years and protect it. These teachings did not lead the people to protect and propagate trees except a few like the peepal, the neem, and the kajiri. Until the beginning of the 19th century human populations were small and forests extensive; forests were considered inexhaustible, invulnerable, and did not need any protection.

British period

With British occupation of the country and the beginning of commercial exploitation of forests to meet the needs of the railways and navy combined with an increase in population, forest areas began to shrink and the quality of forests deteriorated. While the British introduced the scientific management of forests, they alienated the local people and particularly tribals from the traditional use of forests to meet their genuine needs for small timber, fuel wood, and minor forest produce resulting in a reaction against forest and forest administration. Large-scale deforestation and encroachment of forests and tribal revolts were the result.

After India attained independence a National Forest Policy was promulgated in 1952. But commercial exploitation of forests to bring much-needed revenue to the state coffers was continued; further the demand on forests for timber, fuel wood, and industrial needs of pulp and paper increased tremendously. The extension of agriculture also made extensive inroads into forest areas; large areas of forests were destroyed due to submergence under river valley projects.

Post independence period

All these demands on the forests and extensive incursions on forest lands could have been compensated for to some extent had there been commensurate investment in protection of forests from fire and grazing and on plantation forestry on both public and private land. Due to acute resource constraints and lack of realization of the role of forests for the well-being of the land, investment in forests remained a dismal 0.39% to 0.71% of the five-year plan outlays from 1951 to 1985; how could 19% of land area under tremendous pressure be protected, let alone be developed with less than 0.5% of the country's development investment?

Forest lands diverted to other uses

The sad result of these factors has been denudation of extensive areas of forests and a dangerous shrinking of forest boundaries; between 1951 and 1980 4.5 million ha of forest land were diverted to other uses (2.6 million ha to agriculture, 0.5 million ha to river valley projects, 0.134 million ha to industries and townships). Besides the reduction in forest areas due to the above causes, there has been a large-scale deterioration of the remaining forest areas due to the inexorable pressure of populations including tribal populations and their livestock for fuel wood and grazing; this fuel and fodder burden has been for in excess of the carrying capacity of the forests.

Forest estate deteriorates drastically

In the period since 1983, over 1957 million cubic metres of fuel wood has been estimated as removed illicity from the forests, resulting in total destruction of 3 million ha and depletion of 27 million ha of forests. Similarly, with regard to grazing in forests, while the sustainable per annum limit is put at 31 million "cow units" the actual pressure of free and unrestricted grazing comes to 99 million "cow units", an excess of 300% over the permissible limit.

Environmental disaster

All these catastrophic and elemental pressures resulted not only in shrinkage of forest areas and serious depletion in the quality of forests but also in serious environmental problems: soil erosion leading to loss of valuable top soil and annual floods on unprecedented scale resulting in loss of life and property and particularly loss of agricultural crops; droughts leading to reduction in reservoir levels and falls in power generations and industrial production; migration of rural populations to urban areas and consequent social tensions; tribal unrest; and general deterioration in quality of life in urban and rural areas.

Government Initiatives in the 1970s and 1980s

The government realized that forests need protection and higher investments if the gains in developmental programmes are to be safeguarded over the long term. Early in the 1970s the National Commission on Agriculture reviewed the state of forests and forestry in the country and recommended several measures to reduce pressure on forest resources and increase investment in forestry. Some of the important measures suggested are summarized below.

Social forestry

Social forestry/community forestry on a massive scale was recommended to increase availability of wood production; the programme envisaged planting of field bunds along private agricultural lands, shelter belts and windbreaks, plantations on village common lands, road and canal sides, and rejuvenation of degraded forest areas. A massive programme of plantation forestry to meet industrial demand through Forest Development Corporations funded by institutional finance to ensure steady and assured flow of funds was planned. The National Commission on Agricultural also recommended that the existing system of working forests through intermediary contractors should be replaced by forest labour cooperatives to give a stake to forest dwellers in the profits from forest working.

Forest development corporations

In the 1970s almost all states established forest corporations and an area of 500,000 ha had been brought under plantations by these corporations by 1990. Social forestry with international aid from the World Bank, SIDA, CIDA, ODA, US AID, and several others were launched in the decade of the 1980s and over 500 crores of external aid channelled into this programme.

Wildlife Act

The Wildlife (Protection) Act, 1972, was promulgated to protect the flora and fauna of the forests and over 486 national parks and sanctuaries were established covering 4% of the forest area of the country to protect not only the fauna but also promote and preserve biological diversity and indigenous genetic resources of the country.

Forests come under concurrent list of constitution

There was widespread criticism that forest policies pursued by the state governments were not always in the long-term interest of forests. This led to a major development in the fifth plan period (1974-1979) when the Constitution of India was amended to bring the subject of forests under the concurrent list; this gave the central government authority to intervene in the matters of forest working which so far had been the monopoly of the state governments.

During the sixth plan (1980-1985) national policy swung from economic considerations to ecological perceptions. "Development without destruction" was the theme of the Sixth Five-Year Plan; the major thrust of the plan was to save natural forests from the depletion, creation of fuel and fodder reserves to meet local needs, and creation of more national parks and sanctuaries.

Forest Conservation Act

Two major developments marked the decade of the 1980s in forest conservation. The legislative development was the enactment of Forest (Conservation) Act, 1980, which prohibited diversion of forest lands to nonforest use without prior concurrence of the central government and making available alternative lands with adequate funds for growing new forests on such lands. After the enactment only 46,852 ha of forest land diverted to other nonforest use by February 1987 as against an annual average of 140,000 ha during the preceding 30 years.

Separate Ministry of Environment and Forests

The significant administrative development was the formation of a separate Ministry of Environment and Forests in December 1984 followed by similar separate ministries in most states. The creation of such a ministry is an indication of the government's resolve to play a more effective role in the conserving the forest resources of the country.

Revised National Forest Policy of 1988

The National Forest Policy of 1952 was reviewed and revised in 1988. The principle aim of the new policy is to ensure environmental stability and maintenance of ecological balance. The derivation of direct economic benefit must be subordinated to this principal aim. The new policy also laid stress on meeting the genuine needs of the forest inhabitants and particularly tribals by making their needs the first charge on forest produce; it also prescribed promotion of minor forest produce-yielding programmes to improve tribal economies and provide gainful employment to tribals. The new policy prescribed a massive need-based and timebound programme of afforestation as a national imperative.

In order to reduce pressure on forests from industrial demand the policy prescribed wood-based industries should promote plantations on private lands by providing individual growers with inputs including credit and technical and marketing facilities. The new forest policy prescribed "protection, regeneration, and optimum collection of minor forest produce along with marketing of produce" through forest/tribal cooperatives.

Joint forest management initiatives

It has long been realized by several foresters that enlisting the support of local communities in the protection of forests is the most effective way of protecting forest wealth. Pioneers have been making lonely efforts in this direction since the 1970s. This has evolved into the concept of "joint forest management" under which local communities undertake to protect a unit area of forests in return for tangible benefits like a share in the revenue from timber sales and free removal of fuel wood and MFP from designated areas under the overall supervision of the local forester and controlled by a management plan. The scheme has been successful in several experimental locations and is growing from strength to strength in several states; presently nearly one million ha of forests are under joint forest management systems in the country. This will be the main plank of the Eighth Five-Year Plan under the forestry sector and promises immense benefits to local communities and for conservation of forests.

Ecodevelopment

Along with joint forest management systems a massive programme of ecodevelopment of village communities in and around national parks and sanctuaries is proposed in the Eighth Five-Year Plan. The programme envisages an integrated approach to development of the rural forest communities and covers upgradation of cattle combined with improved fodder to reduce pressure of grazing on forests, popularizing biogas and solar and other nonconventional energy sources, wood substitutes, and all such ecologically friendly development initiatives.

Environmental (Protection) Act

Another significant legislative measure taken by the country was the enactment of the Environment (Protection) Act in 1986. The act prescribes nationwide programmes for prevention, control, and abatement of environmental pollution, lays down standards for quality of environment and standards for exhausts and discharge of pollutants from various sources, provides restrictions on location of industries, and provides safeguards for the handling of hazardous substances. The act also prescribes appointment of officers to enforce the provisions of the act.

Coastal area regulations

In 1990 the central government promulgated coastal area regulations prescribing that no developmental activities like mining, industries, and tourism could be located within 500 metres of the high tide line; it also prescribed that fragile ecosystems like mangroves along the coast should be fully protected.

Recycling of waste paper

The installed capacity of paper and paper board manufacturing industries in the country was on the order of thirty lakh metric tonnes in 1989 although actual production is on the order of 19 lakh metric tonnes due to shortages of raw materials. There is tremendous scope for recycling of waste paper. However, a large part of waste paper in used locally as packaging material and for making paper bags, etc. in the small-scale sector; waste paper is urban areas is collected and supplied to the paper mills, which use it to supplement forest raw material resources.

Pilot projects in Bangalore City

Pilot projects to utilize city garbage for production of organic manure (through vermiculture) are being undertaken in some cities, notably Bangalore where the programme is spearheaded by the Centre for Environment Education. It is estimated that Bangalore City can produce 200-300 metric tonnes of organic fertilizers daily from urban garbage besides production of biogas. These programmes help reduce pollution and maintain a clean urban environment on one hand and create employment and organic fertilizers on the other.

Tropical Forestry Action Plan

An unprecedented international initiative launched in 1987, the Tropical Forestry Action Plan, aims at checking uncontrolled destruction of tropical forests while providing sustainable management. It is designed to help member countries develop more fully the broad potential of forestry to provide goods and services for the benefit of the people, particularly rural people, as well as national economies through conservation and wise utilization of tropical forests. The overriding objective of the plan is to assist developing countries in deciding national priorities and in securing the financial support needed to put national plans into action. The Paris summit of the heads of the seven most industrialized countries called for intensified efforts in implementing TFAP; 85 Ministers of Agriculture from 150 countries gathered in Rome for the 25th session of the FAO Conference expressed their strong endorsement of the TFAP initiative. India has drawn up a National Forestry Action Plan and has sought international aid from the World Bank and other donor agencies for its implementation during the decade of the 1990s.

Conclusions

To sum up, forest conservation in India has come full circle. It started at the dawn of Indian civilization when the communities had full rights over all the forests in and around settlements when populations were small, life styles simple, and the demands on forests were few; they lived in hormony with the forests and natural resources around them.

The advent of British, the introduction of commercial forestry, the phenomenal increase in populations, and the heavy demands of industrial civilization on natural resources, particularly forests, lead to a progressive diminution of the rights and concessions of local communities on one hand and depletion of forest resource on the other.

Recent legislative and administrative measures and management initiatives aim at reducing the dependence of people and industries on natural forests and restoring the rightful place of local communities in the management of forests. These measures in the decade of the 1990s will see the restoration of our forests to their original status, stature and extent. Indians are confident that we will enter the 21st century with a track record on forest conservation of which we will be justly proud.

COUNTRY PAPER: INDONESIA

Affendi Anwar
Chairman
Regional and Rural
Development Studies
Program
Graduate School
Bogor Agricultural
University
Bogor, Indonesia

1. Introduction

The Republic of Indonesia is a large archipelago consisting of more than 13,000 islands, of which about 970 are inhabited. The republic lies between mainland Asia and Australia and straddles the equator for some 5,000 km from the Indian Ocean to the Pacific. Five important islands or parts of islands account for some 85% of the land area. These are Sumatra, JavaBali, Kalimantan, Sulawesi, and Irian Jaya. It is estimated that the total land area of Indonesia is approximately 191 million ha. The population of the country rose from 97 million people in 1961 to some 179 million in 1990 and is projected to reach nearly 200 million by 1995.

Since historical times, people have concentrated on the so-called Inner Islands of Java and Bali with their fertile volcanic soils, while the population density in the Outer Islands is very low. Hence, unbalanced population distribution in Indonesia has become a pressing environmental problem.

Intensive cultivation under high population densities in Java and Bali has been practiced for centuries. The introduction of estate crops during colonial rule in the 19th century contributed to the depletion of some natural forests in these islands, and growing population pressure has induced erosion and the expansion of critical lands. These Inner Islands comprise only 7% of the total land area of Indonesia, but are inhabited by more than 60% of the population and have produced some two-thirds of the country's food supply in recent years.

The high population of the Inner Islands which has reached a density of an average 800 people per km^2 has led to a programme of resettlement into the Outer Islands since the beginning of this century. The programme is still actively pursued under the auspices of the Ministry of Transmigration. It appears that the island of Sumatra, which was the earliest target of transmigration, as well as Sulawesi are approaching their saturation points. The main thrust of the transmigration programme is therefore now directed toward Kalimantan and Irian Jaya.

Although historically famous for their spices and gold, the generally poorer soils of the Outer Islands have not invited large settlements in the past. Plantation crops such as coffee, cloves, cinnamon, rubber, and coconut have been established there since the 19th century and today have become a major source of revenue. But the largest part of these islands is covered by tropical rainforest which is claimed to be the second largest tropical forest area within a state after the Amazon forest in Brazil.

Timber and other forest products have been exported from the Outer Islands for centuries. However, when the country faced development resource scarcity in the early stage of development (in the late 1960s), forest resources were considered available natural capital to be exploited. In order to overcome shortages of capital, technology, and skilled and experienced manpower to realize the great value of natural forest for industry and trade, the Indonesian government at that time invited foreign investors to form joint ventures with domestic companies. Foreign investors are interested in logging operations due to greater economic incentives such as tax holidays. Since then forest concessions have become a basis for forestry operations and forest resources have been tapped on a large scale as an impetus to national development. Timber export has since been the second largest foreign exchange earner after oil, first in the form of logs and after a transition period from 1981 to 1987 as sawn wood and plywood.

According to available information, the annual volume of timber exports remained fairly constant between 5 million and 5.9 million tons in the period since 1981. However, the value has more than doubled from less than US$1,000 million in the early 1980s to over $2,000 million in 1987. This corresponds to an increase in its share within all nonoil export commodities from 20% to 24% and within those of the agricultural sector from 32% to 41%. Helped by a decline in revenue from the oil sector, the earnings from timber exports in 1986 and 1987 were well over 60% of the net revenue from oil and natural gas, a remarkable increase from the 10% in 1981.

2.   Information On Forest Resources

Among tropical rainforests, Indonesia's forests are of regional and global importance. Indonesia is thought to have nearly 60% of all tropical forest in Asia and perhaps 90% of the remaining virgin stands. Indonesia has the world's richest forest in terms of commercial production. East Kalimantan, for example, has the most homogeneous and most valuable dioterocarp forests in the region. Forest products are Indonesia's most important nonoil export commodity, and they provided about $2.5 billion in foreign exchange in 1987. Indonesia's forests also provide poles and timber for domestic construction, and minor forest products such as rattan, resin, turpentine, and bamboo make an important contribution to the economy. Indonesia's rainforests are also a reserve of flora and fauna of potential economic value. Tropical forests in general include perhaps one-third of the earth's total species, and the value of only a small fraction is known. Finally, the value of these wild species of plants and animals become important as contributions or sources of domesticates, as sources for development of new pharmaceuticals, or as sources of genetic stock for the multibillion dollar international agribusiness industry, which are rarely estimated.

As Indonesia is richly endowed by natural forests, more than half of the land area equal to 191 million ha is covered by forest. According to land use data presented in the so-called "Compromised Land Use Data" (*Tata Guna Hutan Kesepakatan (THGK)*) there are 30 million ha of protected forest, 19 million ha of nature preserves, 31 million ha limited production forest, 34 million ha of permanent production forest, and 30 million ha of production forest that can be converted to other uses.

The size of forests in Indonesia, however, differs by estimating source. The FAO, for example, estimated from a mapping survey of 1981/82 for Sumatra, Kalimantan, and Irian Jaya that the total forest cover in Indonesia is only 114 million hectares, while the Regional Physical Planning Program for Transmigration (RePPProT) which was analyzed by Land Resource Development (LRD) under the British Overseas Development Authority (ODA) estimated the forest cover reaches 116 million ha (Table 1). The FAO data as well as that of the ODA did not include the forest damaged by fire in Kalimantan in 1983, which is estimated at 3.6 million ha, and this data did not include the amount of newly harvested forest.

Table 1. <u>Estimated forest cover in Indonesia (ha)</u>.

Island	FAO estimate	ODA estimate
Sumatra	22,200,000	23,300,000
Kalimantan	35,400,000	39,600,000
Irian Jaya	38,000,000	35,000,000
Other	18,000,000	18,000,000
Total	113,600,000	115,900,000

Source: World Bank, 1989.

Based on those two factors, it is estimated that the forest cover remaining at present is only 110 million ha. In this remaining forest are the resources that can generate many benefits, including timber and nontimber products and can generate many public or social services such as climatic, soil, and watershed protection, as well as amenities that are demanded in national development. However, the land covered by forest is only 60% of the land mass of Indonesia. Further, the forest land is subdivided into forest types according to function, as depicted in Table 2.

Table 2. **Estimated forest by type**.

Forest type	Ha	
Production forest	49.3	million
Protection and conservation	39.9	million
Total forest cover	118.0	million
Total land area:		
Outer Islands	177	million
Java and Bali	14	million
Total land area	191	million
Land suitable for agriculture	63	million

Source: FAO Report, 1990.

According to a report by the World Bank, it is important to note that this is indeed a very large forest area. For comparison, Indonesia, which is roughly the size of Western Europe (excluding Scandinavia), has 2.5 times as much forest, and it has almost twice as much of its land surface under forest (60%) as the USA (32%). However, while Indonesia is fortunate to have such a large amount of forest resources, there are reasons for concern. The rate of deforestation during the 1970s has doubled from 300 thousand ha/year in 1970 to 600 thousand ha/year in 1981.

The concerns are based on neighbouring countries' experience, which have lost substantial amount of their forest resources in a relatively short period of time. The Philippines and Thailand, for example, were largely forested until this century, and their forests have been reduced to 25% and 30% of the land area, respectively; no more than 3% of land in the Philippines is mature natural forest. Thailand, a major exporter of tropical hardwoods in the 1960s and 1970s, now imports timber for construction purposes, and land degradation associated with deforestation has become so serious in Thailand that all logging was banned in late 1988.

3.   Forest Management

Historical Development of the Forest Management System

There is a long history of forest management in Indonesia. In East and Central Java, intensive management of teak was introduced more than 100 years ago by German foresters for the purpose of successively replacing indigenous teak forest with more productive stands. Plantations of this valuable tree are now extensive and produce large volumes of wood on a fairly sustainable basis.

During the earlier part of the twentieth century, plantation management was extended to a number of other species, in Sumatra mainly, using conifers and the hardwood roses, eucalyptus, and mahagoni. More recently, fast-growing plants such as Callahan and Leucaena have been planted for fuel wood and conservation purposes.

By the late 1960s a new phase in Indonesian forestry development commenced and this was stimulated by a very marked increase in worldwide demand for wood for industrial processing. In order to open up the extensive and valuable hardwood resources for trade, the government enacted legislation to encourage the participation in resource development of foreign capital investment, skills, and equipment. The response to governmental initiatives greatly exceeded expectations and led to a boom in forest exploitation. In 1976, some 24 million ha were involved in wood production concessions and today, 15 years later, the area is about 60 million ha. Even before the concession system was introduced, wood was being cut and sold from dipterocarp forests and other forest types. Forest exploitation was very often carried out during the course of agricultural land clearing operations. Until 1969, teak continued to be the most important timber species of Indonesia, but from 1970 on, it was replaced by Meranti (Shorea sp.) in both volume and value.

The concession system provided government with an administrative tool through which systematic control of exploitation could be achieved. By placing specific obligations on concession holders, development objectives such as housing, road building, welfare, and other services and employment could be provided as well as ensuring revenue. Even though forest resources may have seemed almost limitless in 1967 when logging started, there was at that early stage some concern about the sustainablity of wood production and indeed of forest systems generally. No silviculture regulations existed and the only restraint imposed on loggers was a minimum diameter cutting limit. With very little experience within Indonesia in managing tropical high forests, policy was based on a study of Philippine and Malaysian forest management, as well as local experience leading to acceptance and implementation of three separate but not unrelated silvicultural models intended to ensure long-term forest productivity.

The models are known respectively as the Indonesian Selective Cutting System (*Tebang Pilih Indonesia,* or *TPI*), Clearcutting with Natural Regeneration System (*Tebang Habis Pemudaan Alam, or THPA*), and Clearcutting with Planting System (*Tebang Habis Pemudaan Buatan, or THPB*). These systems were to be applied in mixed hardwood forests according to whether these were determined from prelogging assessments to be of high or low productivity, or to be unproductive. These systems, especially *TPI,* continue to provide the framework for extensive management of lowland forests.

Other forest types, notably coastal mangroves, have been harvested on a flexible basis determined by local demand, supply, and accessibility factors. A systematic approach toward their management has not been adopted.

In Search of Systems for Sustainable Wood Production

Indonesian foresters have long recognized and applied in practice the compelling need for permanent reservation of forest land for protecting forests, and regulation of the harvest so that wood can be cut on a long-term, sustainable basis. The management of the teak and pine forests of Java and Sumatra are examples of this experience.

A new challenge and one of much greater dimensions was presented when the question of finding the most appropriate management arrangements for the vast areas of mixed hardwood forest was approached. It was clear that the extensive dipterocarp forests offered scope for yielding large amounts of wood for industrial processing, and that several main types were known from other Asian experience to be easily marketable overseas. Many species were known to have the capacity of producing abundantly through periodic, natural regeneration, thus providing a possible basis for crop management.

Mixed rainforests in Sumatra, Kalimantan, and Sulawesi have been logged since the 1930s, but operations were largely of low impact on the land. Commercial trees greater than 35 cm in diameter were cut and regeneration of the forest was left to nature; there was no silviculture at all. This approach was no longer suitable when large-scale mechanized and semi mechanized logging commenced in 1967. The wide extent of mixed hardwood forests, the strong interest in introducing log production in several provinces simultaneously, limited access to forests, limited numbers of skilled and experienced foresters, and only limited silvicultural knowledge of Indonesian dipterocarp forests suggested that forest management needed to be approached extensively, in contrast to the intensive management of teak, pine, Agathis, and other plantations.

In 1966, the Directorate of Forest Utilization instructed that logging must be undertaken on a selective cutting basis with a rotation of 60 years. It was the first government regulation in natural forest management. Of underlying concern was the long-term goal of achieving sustainability of forest systems and future wood production. A year later, General Guidelines for Forest Exploitation were issued in an Appendix to a Decree of the Minister of Agriculture. These provided scope for either selective cutting or clear felling of forests and prescribed that stand and regeneration inventories be conducted one year prior to logging in each operational unit. In a move to tighten cutting and postlogging practices further, regulation of logging operations was introduced in 1970. The Government Regulation No. 21/1970 states that "the forest exploitation right was only given on the basis of selective forest ... with the obligation to conduct regeneration naturally or artificially (planting), and to tend the forest." Although the regulation is well intentioned, it was quickly found that application on the ground of this approach was unsatisfactory and led to an unacceptable amount of damage to the residual forest. There was also a recognition of considerable variation in species composition and forest structure, yields, and geography in all parts of Indonesia between Sumatra in the west and Irian Jaya in the east. A wider framework for rainforest management was needed which would strengthen the government's aim of not only achieving industrial log production but also providing it with the confidence that the essential elements of ecological structure and composition would be preserved to enable further harvesting to be made in the future.

At the request of the Director General of Forestry in the late 1960s, a team of specialists of the Forest Research Institute, Bogor, studied all available local knowledge and experience at the time. The team also reviewed progress in dipterocarp management in Malaysia and the Philippines. In 1970 the team proposed the use of two silvacultural systems based upon replanting. The three systems were issued with the Decree of the Director General of Forestry No. 35/1972 and are now widely known as the *TPI*, *THPA*, and *THPB* systems.

Thus, a substantial step forward in the search for systems for sustainable wood production was made and have served as the backbone of forest management for more than 15 years. by far the most widely known and applied has been the Indonesian Selective Felling System or *TPI*, which was designed for mixed dipterocarp forests having a normal range of size classes found in the forest. The Clearcutting with Natural Regeneration System, <u>THPA</u>, is more appropriate if the forest is dominated by only one or a small number of dipterocarps, the stem diameter distribution is skewed to the larger-sized trees, and seedling regeneration is plentiful. The Clearfelling with Replanting System was designed primarily to convert poorly stocked forest and poor secondary forest into more productive forest plantations.

Recent Issues Influencing Forest Management

Although the Indonesian Selective Felling and other systems provided a very much firmer management base than had existed previously, it is useful to recognize that a number of changes, or new influences, have emerged over the last 10 to 15 years having a bearing on present and in all likelihood future management perspectives. The more important are as follows.

1) The burgeoning demand for industrial wood in local as well as overseas markets coupled with a wider wood market and the acceptance of species formerly considered to be of no commercial value -- the so-called lesser-known species -- has been a notable trend. The management impact has been to cause the removal of more wood from some forests and to place greater pressure upon the forest environment to recover and yield wood in the future.

2) Production from forests has increased in importance as government has encouraged development of nonoil natural resources through out Indonesia. The forestry sector is prominent in generation of revenue from exports and it can be expected to continue to be prominent in national economic planning and business investment. Two clear implications for

management are first, to strengthen the arguments for protection of the permanent forest estate and to do so with dedication and care, and second, not to weaken in the application of the principle of sustainability. To give way in respect of these two points could seriously undermine the broad government policy of achieving economic diversification.

3) Forests provide significant developmental benefits regionally, especially through employment in forests and mills. To sustain present employment and to provide for expanded employment opportunities in the future implies, as mentioned in 2) above, the compelling and urgent need to conserve the permanent rainforests and expand the forest estate of high-performance plantations. Also implied by the wider social value of forests is a need for thoughtful planning and of effective plan implementation directed toward meeting clear, attainable, and balanced objectives.

4) A fourth trend concerns the wider knowledge of forest structure and growth and of the behaviour of forests when manipulated silviculturally. Research into flowering and seeding patterns, seedling growth and plant propagation from cuttings, growth responses following cleaning and thinning, and the use of other silvicultural techniques has proceeded steadily since the 1970s. Much of the work has been aimed at evaluation of the effects of *TPI* implementation in various forest types. Results of research have been important in strengthening the confidence of foresters technically and in providing scope for more intensive management.

5) Last, but by no means least, is the growing awareness and significance of a range of social, biological, and physical environmental values of tropical rainforests. Environmental issues will doubtless influence management differently from one locality to another but as is being experienced elsewhere, including in several other Asian countries, it can be anticipated that issues of the wider environment in which mankind and forests co-exist will have an increasing bearing on forest policy development and management practice. The following six points are likely to be important: water conservation and control of soil erosion; the supportive or complementary value to agriculture; a resource (incompletely known presently) from which a range of food plants, nonwood products, and plants having medicinal and chemically significant products can be harvested; the scientific significance of complex ecosystems providing habitat for a large array of unusual, and in some cases, unique species, the genetic diversity of both being incompletely known to science; a possible worldwide influence of forests, especially in tropical regions, on climatic change; and conservation of the productive potential of tropical forests

and forestland from local and external economic demands for wood products and from the competing demands from other sectors for access to forest land.

Identification of the more prominent issues and trends which have emerged since the 1970s will be helpful later and should be studied further, particularly in consideration of more confident management of the forests in Indonesia.

4. Forest Degradation and Environmental-economic Problems

Forest degradation has been an important global issue recently. It is especially important considering the linkages of deforestration to changing weather conditions, soil degradation, decreasing agricultural productivity, loss of wildlife and their habitats, and depletion of biodiversity.

Recent study shows that there is also evidence that Indonesia's forest area is being rapidly reduced. In the early 1970s, the FAO estimated the rate of deforestation in Indonesia at about 300,000 ha/year; in 1981, this estimate was raised to 600,000 ha/year; and recently, the FAO suggested that it could be as high as 1 million ha/year. How rapidly Indonesia's forests are disappearing will not be known until reliable time-series data are available. However, figures from several sources tend to confirm the FAO estimate. The World Bank, for example, has reasonable figures on forest conversion for government-sponsored tree crop and transmigration programmes, and this is estimated at about 200,000-300,000 ha/year in the third five-year plan (Repelita III, 1979-1984).

In recent studies of wood processing, the consulting firm INPROMA from Atlanta, USA, estimated deforestation due to logging at about 80,000 ha/year, or about 10% of the annual area logged; destruction due to forest fires was estimated at about 70,000 ha/year. The latter figure does not include the Kalimantan fire and may be on the low side.

The greatest uncertainties relate to the loss of forest due to shifting cultivation and smallholder agricultural conversion outside of development projects. Recent RePPProT studies indicate that 14 million ha in Sumatra, 11 million ha in Kalimantan, and at least 2 million ha in Irian Jaya are under shifting cultivation or under brush and secondary forest which usually signify previous agricultural use. If this area is expanding at only 12% annually (about half the rate of population growth in the provinces with large areas under shifting cultivation), then deforestation due to various types of smallholder forest conversion in the Outer Islands would be on the order of 500,000 ha/year. The figure could well be higher. Of this, perhaps half could ultimately regenerate as secondary forest, to be cleared again by smallholders when fertility improves, but virtually all would be lost to timber production.

The figures depicted in Table 3 suggest that deforestation during 1979-1984 could well have approached 900,000-1,000,000 ha/year. The data are rough, but any reasonable range (e.g., deforestation of 700,000 to 1.2

million ha/year) is significant. It is difficult to disaggregate these figures since logging contributes to deforestation by developing roads, which open new land to smallholders, and by providing wage work, which attracts families into the forest. With these caveats in mind, however, these figures suggest that of the area deforested, smallholder agricultural conversion may account for about half, development projects about one quarter, and the remainder due largely to poor logging practices and forest fires. Table 4 shows average rates of deforestation by island.

Table 3. **Souces of deforestation (ha/year).**

Source	Best estimate	Range
Smallholder conversion	500,000	350,000-650,000
Development projects	250,000	200,000-300,000
Logging	80,000	80,000-150,000
Fire loss	70,000	70,000-100,000
Total	900,000	700,000-1,200,000

Source: World Bank calculations

Table 4. **Average rates of deforestation by island**.

Province	Period	Loss of forest cover (as a percentage fall in proportion to total land area)
Sumatra	1950-1984	- 0.67
Kalimantan	1950-1982	- 0.37
Sulawesi	1950-1982	- 0.45
Maluku	1950-1982	- 0.10
Irian Jaya	1950-1982	- 0.52
Outer Islands except NTT		- 0.44

Another analysis has shown strong correlation of deforestation (for various purposes) with population density, agricultural productivity, and growth of real income. An increase in population density of one person per square kilometre reduces forest cover by 0.8%. An increase of 100 kilograms per ha per crop in wet rice yield (a proxy for agricultural productivity) increases forest cover by 4.6%. An increase in income per capita of 1,000 rupiah per person increases forest cover by 0.015%. The cumulative effect of unexplained factors including accelerated development of logging (and other) roads during the last 20 years has been an average decline of 3.7% per year.

Long-term trends in deforestation based on inventory data covering the period 1950-1982/1984, gives a compounded rate (of forest loss) of 0.44% per annum. However, in recent years the rate of forest loss has shown a higher rate, partly influenced by the Kalimantan fire.

Natural forests as at the beginning of 1990 have been estimated to be about 109 million ha with reduction in all categories of forest. By 2050 it is predicted that about one-fourth of the forest cover remaining in 1990 will be lost. Most of it accounted for by loss of conversion forests and production forests (assuming that protected and conservation forest will remain intact).

The impact of deforestation in one sense incurs some economic cost. Simple analysis by the World Bank shows that deforestation at the above level is associated with very high costs. One ha of standing timber in primary forest has a net present value (NPV) of at least $2,500-3,000, and about $500 if already logged. Shifting cultivation, yielding 2,000 kg of rice per ha for one year with a fallow period of 15-20 years, produces an NPV per ha of about $120. Assuming 50% of the area opened is in loggable forest and the rest in logged-over areas, the net loss to the economy would be conservatively $625-750 million/year. With another $150 million lost to logging damage and fire, losses would be about $800 million. Loss of minor forest products could bring this figure to $1 billion/year. The additional loss of timber on sites cleared for development projects could be $40-100 million, although this would ordinarily be offset by agricultural benefits of greater value.

In addition to being costly to the economy in terms of foregone timber production, deforestation has a number of negative consequences of national and international concern. Large-scale clearing for agricultural produces smoke and carbon dioxide which contribute to global warming; most studies place Indonesia second only to Brazil in producing atmospheric pollution from this source. Forest clearing also threatens biological diversity and endangered species. This is particularly serious in Indonesia where its island topography results in a high degree of *endemism*, and isolated species can be quickly eradicated if the forest cover is destroyed.

At the national level, deforestation also jeopardizes Indonesia's economic objectives. In particular, it threatens the wood raw material supply on which export diversification partly depends. This leads to land degradation that disrupts regular water supplies and reduces the productivity of both traditional cultivators and downstream water users. This following section focuses mainly on issues related to wood raw material supply and on the policies needed to realize the economic objectives of Indonesia's development planners in the timber sector.

5.   The Role of the Forest in Development

About 144 million ha or 75% of Indonesia's land falls within Forestry Department boundaries. The area is divided into five categories: forest set aside for conservation and national parks (13%); forest intended primarily for watershed protection (21%); limited production forest (21%); and regular production forest (24%), which can be converted to agriculture and other uses. Data by province are given in Table 5. Of the 144 million ha, about 113 million ha are within permanent forest categories, of which about 65 million ha are in limited and regular production forest. However, not all of this land is forested.

In order to evaluate the area under closed canopy, forest within those areas classified as reserves, protected, and production forest, the RePPProT team, at the request of the World Bank, superimposed Forestry Department boundaries on 1981/1982 aerial photographs of Sumatra, Kalimantan, and Irian Jaya. Table 6 shows the percentage of land **not** under closed canopy forest in 1981/1982. Within the areas set aside for conservation and protection, about 16% of the land in Sumatra and 8% in Kalimantan was deforested; within the area classified as limited production and production forest, about 30% in Sumatra and 16% in Kalimantan had been converted to other uses. On a provincial basis the figures were even more significant. Sixty percent of the permanent forest area in Lampung was deforested, 43% in South Sumatra, 42% in North Sumatra, and 44% of limited production forest in West Kalimantan was gone. These provinces with high levels of deforestation within Forestry Department boundaries are generally provinces in which significant development is taking place, and given the widespread agricultural development in the Outer Islands since 1981/1982, these figures could now be higher. This amount of deforestation has clear implications for wood raw material supply and for sustainable rates of timber production.

Table 5. Area within forest boundary by Forestry Department classification (1,000 ha).

Island	Reserve	Protection	Limited Production	Production	Conversion	Total
Sumatra	3,684	7,094	7,579	6,821	5,032	30,210
Java	444	554	0	2,014	0	3,012
Kalimantan	4,101	11,415	14,234	8,293	44,967	
Sulawesi	1,406	3,867	3,926	2,092	1,993	13,284
Irian Jaya	8,312	4,732	7,123	11,775	40,591	
Other	779	3,229	2,874	1,581	3,444	11,907
Total	18,726	30,317	30,526	33,865	30,537	143,971
(%)	13	21	21	24	21	100

6,924 | | | | | | |
6,924 | | | | | | |

Source: Department of Forestry, 1986/1987.

Table 6. Percentage of area within Forestry Department boundaries not under closed canopy forest.

Province	Reserves	Protection forest	Limited production forest	Regular production forest	Production forest
			— % —		
Sumatra					
D.I. Aceh	3	5	20	26	49
North Sumatra	4	45	46	37	65
West Sumatra	7	27	29	36	38
Riau	15	37	24	8	58
South Sumatra	37	50	49	42	65
Jambi	13	12	13	17	42
Bengkulu	8	12	18	29	45
Lampung	33	76	none	72	85
Sumatra subtotal	**16**	**33**	**30**	**29**	**58**
Kalimantan					
West Kalimantan	7	13	44	23	39
Central Kalimantan	25	5	7	18	47
South Kalimantan	60	35	18	34	59
East Kalimantan	4	1	1	7	22
Kalimantan subtotal	**9**	**8**	**14**	**17**	**36**
Irian Jaya	**15**	**13**	**9**	**9**	**19**
Total	**14**	**17**	**18**	**18**	**37**

Source: LRD/RePPProT Studies, 1986, 1987

Ecological issues related to forestry are gaining attention. Conservation of the environment has protection links with economic development. The contributions of forests to watershed protection are recognized. Protected forests have been established across the country.

Other forest types such as mangrove forests are critical for sustaining maritime ecosystems and fisheries resources because they constitute a food chain for many on-shore and off-shore marine resources. Indonesian tropical moist forests are a depository of an incredible wealth of animal and plant species. For these reasons, Indonesian forest lands have been set aside as parks and conservation areas to serve many of the environmental roles that are gaining in importance.

In the past, such allocation of forest areas was not considered economical, but wasteful because of the underdeveloped forest produced neither lumber nor agricultural products. However, now it is more and more recognized that such "idle" reserves have many economic as well as ecological functions, such as non timber products that can be managed sustainably and give benefits to both local and regional people, either as consumption items or as sources of cash.

In relation to reducing rural/urban and regional disparities, which is another important national goal, the forest plays important roles. In many rural areas, forestry and forest products are the main source of cash income. In these underdeveloped rural areas, forestry is called upon to become the main vehicle for economic development.

The economic benefits of forest conservation have recently been acknowledged by Indonesian policy makers. Indonesian forestry has now entered a period in which forests are no longer considered an unlimited or inexhaustible resource. They are now perceived as a source of many benefits. They have gained importance in national economic development, due mostly to the export diversification and job creation strategies, and the goal of reducing regional disparities in income and development.

Forests are a limited renewable resource. Policy makers have acknowledged the limits of the forest resources available and the need for sustainable management of this natural endowment. Better and more rational utilization of natural forest resources and development of forest plantation resources are recognized national goals. Forest management is no longer seen as only a timber-oriented activity. It is concerned with multiple purposes of forest production (wood and nonwood), conservation, and environmental protection. There are still several imbalances and deficiencies in translating policies and strategies into action. This has affected the development of the sector in the desired fashion. This has also resulted in management stress and resource depletion. Many of the problems involved are institutional in nature.

Providing timber raw materials to an expanding export-oriented economy from a shrinking resource endowment is a complex issue. This is further aggravated by the need also to supply a large number of other products to provide environmental, conservation, and protection services, and to ensure sustainability of production for the benefit of future generations. Alternative economic sources of timber supply must be found in order to satisfy expanding markets and to reduce the pressures on the resources. High-yield timber plantation programmes are being promoted by the government. Ensuring private involvement in these programmes is a major goal that requires an adequate package of incentives and policies.

However, large-scale plantation projects are not in themselves sufficient to address the needs of the local population. Hence, agroforestry systems and development of home gardens and small private forest/woodlots, which has so far been practiced mostly in Java, needs to be disseminated to the Outer Islands, especially around human settlement areas. More effort needs to be made to promote utilization of the timber resources available in tree crop estates of rubber, oil palm, and coconut.

The prevailing shifting cultivation practices need rational utilization. There are three types of shifting cultivator. One is traditional and more professional who knows how to maintain the sustainablity of the forest. The other is the opportunistic shifting cultivator who is often a former labourer in a logging operation and who turns to shifting cultivation, but lacks knowledge of traditional techniques of sustainable shifting cultivation. The third type is the commercial shifting cultivator who is associated with a trader and markets the produce coordinated by a city businessman. Supporting shifting cultivators of the first two types with technically sustainable systems, and providing alternative sources of cash income, are two of the actions needed to reduce forest degradation caused by slash-and-burn agricultural practices. In this context development of agroforestry solutions, buffer zones to protect the tropical forests, and promotion of cottage industry gain relevance. However, the last type of shifting cultivator must be banned, because it is too commercialized and it is difficult to control.

Indonesia's forest management model for the natural forests, the *TPI* or the Indonesian Selective Cutting System, although sound and attractive from a conceptual point of view, has generally failed in the field due to difficulties (high transaction cost) in enforcing the regulations. Modifications of the system are required in order to ensure sustainability, to reduce waste, to increase productivity, and to facilitate implementation. The system should be flexible enough for managing diverse forest resources.

Rationalizing the allocation of forest land for purposes of protection and conservation, and ensuring that forest exploitation and management are carried out taking environmental values into consideration, will require carefully designed policy packages and regulations. Systems of integrated, multi-purpose forest management need to be developed to realize its full potential. This in many cases requires active market promotion and extension support, particularly in relation to less traditional products.

Indonesia has adopted a model of public ownership and private utilization of the natural forests. Management of and investments in the forestry sector will necessarily continue to be based on the private sector. Not only does the large size of the sector forbid models of public management, but a reduced public-sector role is also a major target of the Indonesian government. This raises complex problems, because many of the values of forests are not internalized by private companies, and policy incentive packages must be devised to mobilize the private sector and to ensure optimum management and investment in the sector. Policy packages to ensure management and exploitation of the natural forest by the private sector in a desirable fashion are often difficult to implement. Concession policies, such as size of concessions, duration, evaluation criteria, renewal of concessions, royalty and other payments, infrastructure development, logging plans, overlapping rights, and so on may need to be modified to insure that practices conform to the government's development goals.

Public ownership and private use raises the issue of how to ensure the capture of rents generated by forest activities that are due to the owners of the resource, the public. The royalty system in place has failed to capture economic rents. It has also contributed to wasteful practices in logging and processing. Furthermore, since many of the benefits of forests are not priced, the resource tends to be undervalued in the overall design of forest revenue systems. If the forest resource is to be conserved for posterity, this aspect needs to be addressed on a priority basis. The forest revenue system will also have to be reassessed in the light of the increasing importance of plantation forestry and the need to ensure remunerative prices for plantation wood.

6.   Forest Plantation and Afforestation

Forest regeneration through plantations was first initiated in Java in 1980 using teak (*Tectona grandis*). During the First World War the plantation forest was established in Sumatra using pine (Pinus merkusii). Plantation work was continued between the two world wars on both islands as well as in Sulawesi and Nusa Tenggara. The range of species was extended to a number of hardwood species, and more recently also to indigenous and exotic fast-growing species. Some 80% of forest plantation work, especially in Java, is carried out through the Tumpansari system which is a <u>Taungya</u> system. Some 140,000 persons are estimated to be engaged on a part-time basis in this work.

Since the introduction of the Five-Year Development Plans in 1969/1970, the plantation programme has been increased significantly on all islands except Maluku and Irian Jaya.

<u>Types of Forest Plantations</u>

The following plantation activities are monitored or carried out under the direction of the Directorate General of Reafforestation and Land Rehabilitation:

1)   plantations for timber production and rehabilitation within forest land of Java, under the responsibility of the State Forest Enterprise Perum Perhutani;

2)   rehabilitation planting (reafforestation) within forest land of all other islands except Java, under the supervision of the Directorate of reafforestation and Regreening.Subdirectorate of Reafforestation;

3)   rehabilitation planting (afforestation) outside forest land of all islands including Java, under the supervision of the Directorate of Reafforestation and Regreening, Subdirectorate of Regreening , jointly with the Directorate of Fodder Crop Development of the Ministry of Agriculture, the Ministry of Public Works, and the Ministry of Home Affairs (Sekretariat Batuan Penghijauan dan Reboisasi Pusat((rehabilitation work is assisted by the Directorate of Soil Conservation whose responsibility is to construct dams, check-dams, and terraces);

4)   plantations for timber production outside Java, mainly by forest concessions, under the supervision of the Directorate of Industrial Timber Estate (*HTI*);

5)   buffer zone development for people surrounding production and conservation forest, agroforestry activities, and minor forest product development, all under the supervision of the Directorate of Reafforestation and Regreening, Subdirectorate of Social Forestry and Silviculture, with various links with different government agencies, including the Ministry of Transmigration and the Directorate of Forest Extension, Directorate of Forest Protection and Nature Conservation, Perum Perhutani for agroforestry activities in Java, and agroforestry projects assisted by foreign donors; and

6)   private plantations and so-called nucleus estate schemes called the Sengonisasi Programme (afforestation using <u>Albizia</u> sp.) in Java have only started two years ago with the objective of providing raw material for the pulp and paper industry.

Not included in the list is enrichment planting under the Directorate General of Forest Utilization, which is regarded as silvicultural activity within natural forest.

Forest Plantation in Perum Perhutani of Java

The responsibility of managing timber estates on Java is vested in the State Forest Enterprise Perum Perhutani. The enterprise controls a forest area of more than 2 million ha of which 57% are forest plantations. Data and information related as to the extent of forest plantation are depicted in the Table 7.

The area of 1,884,504 ha is actually and potentially productive plantation area including also forest and non-forest with the potential of being converted to forest plantations. Some of these areas consist of plantations that are not suitable for the main species of the respective working cycle. Excluding mangroves (Rizophora), some 73% of the plantation area is actual plantations; of this 84% is suitable for the main species; and of this again only 69% is managed under a clear-cutting regimen. The rest are plantations of poor quality being managed under selective cutting.

Growth and Yield

Estimates of growth rates have been made for various species and plantation sites. There exist also yield tables for teak on Java. The mean annual increment at rotation age for the main species is usually given as in Table 8. There are, however, few records on actual yield. Even for well-managed teak, the records appear very much incomplete. When the reported final yield per ha is divided by the rotation, average values of 1.15-2.43 m 3 ha/year are obtained. These values do not include intermediate thinnings but they are nevertheless very low and reflect the immense pressure on wood resources on Java rather than a poor potential of growth.

Rehabilitation within Forest Land Outside Java

The records under this category are said to include plantations established by concessions and the State Forest Enterprises *Inhutani I, II, and III* prior to the Industrial Timber Estate Programme *HTI*. Part of the plantations is therefore for production and another part for protection and soil conservation; the records about the functions of these plantations are incomplete.

The difficulties in maintaining these plantations, especially those for protection and soil conservation, are due to low survival rates. In an attempt to obtain some indication of the extent of the plantations and their species composition, the survival percentages have been used to express so-called successful plantations in area equivalents, and records of species planted were used for the breakdown by species and species group. Provincial details of this breakdown are given in a summary by region in Table 9.

Table 7. **Extent and conditions of forest plantations under Perum Perhutani, Java, October 1988 (ha).**

Species of working cycle	Area actually & potentially productive	Actual area of forest plantation	Actually suitable for main species	Under clear-cutting regimen
Tectona	1,080,868	880,508	774,716	572,937
Pinus	410,693	282,819	231,487	147,717
Altingia	95,943	34,991	32,866	17,087
Agathis	73,679	45,393	35,128	12,431
Swietenia	35,143	31,605	12,457	4,761
Dalbergia	22,518	15,700	4,531	1,331
Melaleuca	6,584	6,490	6,328	4,866
Other	126,555	58,146	35,438	18,193
Sum	1,851,983	1,355,652	1,132,951	77,323
Rizophora	32,521	8,687	8,681	6,972
Total	1,884,504	1,364,339	1,141,632	786,295

Table 8. **Mean annual increment (MAI) and length of rotation cycle for respective species.**

Species	M.A.I. m^3/ha/year	Rotation assessment in years
Tectona grandis	5-12	40-80
Swietenia macrophyla	14.8	at year 35
Albizzia falcataria	40-50	12-15
Sesbania grandiflora	20-25	5-10
Eucalyptus sp.	20	20
Eucalyptus deglupta	26	at year 5
Acacia auriculiformis	23	at year 8
Acacia mangium	26	at year 8
Pinus merkusii	15-18	25-30
Agathis loranthifolia	20	50

Table 9. **Rehabilitated area within forest land up to 1988 fiscal year by region and species group (Java excluded)***

Region	Area Planted	Area equivalent of successful plantations					
		Pinus sp.	Agathis sp.	Slow hardwoods	Fast hardwoods	Mixed species	Total
Area in ha							
Sumatra	493,580	153,534	117	87,155	28,632	72,092	341,530
Nusa Tenggara	124,654	296	-	28,084	13,874	29,746	72,000
Kalimantan	205,772	46,097	236	7,899	6,397	20,221	80,850
Sulawesi	395,893	110,798	327	26,025	26,063	37,257	200,470
Maluku	1,915	-	-10	275	135	-	420
Irian Jaya	-	-	-	-	-	-	-
Total	1,221,614	310,725	600	149,438	75,101	159,316	695,270
Percentage of species							
Sumatra		45	0	26	8	21	100
Nusa Tenggara		0	0	39	20	41	100
Kalimantan		57	0	10	8	25	100
Sulawesi		55	0	13	13	19	100
Muluku		-	2	66	32	-	100
Irian Jaya		-	-	-	-	-	-
Total		45	0	21	11	23	100

* Indicative only.

The area and species composition in the tables should be regarded as indicative only: the species records used go only to 1983 with no allowance for reported or actual failures. Damage or loss after the survival assessment, usually three years after planting, is not taken into account, and some of the plantations for timber production have now been harvested and may or may not have been replanted. With these reservations in mind, the actual area planted in this category amounts to approximately 1,200,000 ha of which an equivalent of some 690,000 ha may be established plantations. Nearly half of them, or 45%, are likely to be pine, 21% slow-growing hardwoods, 11% fast-growing hardwoods, and 23% a mixture of various species. According to these data, the total remaining critical area within forest land outside of Java is estimated to be between 5 to 6 million ha, where the first figure is the sum of critical land according to Forestry Statistics plus the failed plantations, and the second figure is the estimate by the Subdirectorate of Reafforestation.

According to the Directorate General, the target during the Fifth Five-Year Development Plan is to rehabilitate 4,900,000 ha that are nearly all critical areas within forest land. However, the area of plantations destroyed again each year are considerable. It was reported in 1989, for example, that at least 470 ha of trees planted in Sumatra Barat under the reafforestation programme have been destroyed by fires during that year's dry season. Some of the fires may be accidental but many were probably set by hunters as is customary practice. In addition to an effective system of fire protection it is necessary to educate the people in the benefit of reafforestation through increased extension work and to bring about a stop to such deliberate devastation.

Rehabilitation outside Forest Land

Until 1978, critical areas outside forest land were rehabilitated by establishing vegetation cover only, mostly with forest, fodder, and fruit trees. Since the Third Five-Year Development Plan, rehabilitation work has been assisted by the construction of checkdams and terraces where necessary. The total area reported as rehabilitated amounts to some 5,800,000 ha; details by province and year are given in Table 10. The survival rates in this category of plantation are reported as 27-71% in different years, with an approximate average of 50-60%.

The remaining critical area stated by the three sources listed in Table 10 agree quite well in the order of magnitude, with the exception of two apparent errors in the data. Again the estimates of the Directorate of Reafforestation and Regreening are larger in some provinces than those of forestry statistics and of the Directorate of Fodder Crop Development. The extent of critical areas outside of forest land at the beginning of the Fifth Five-Year Development Plan, i.e., April 1989, was estimated to be between 5.7-7.3 million ha, where the first figure is the estimate of the Directorate of Fodder Crop Development with the area of Kalimantan Timur adjusted to 4,000 ha, and the second figure is the estimate of the Directorate of Reafforestation and Regreening. The planted areas and the remaining critical areas are summarized by region in Table 10.

Table 10. **Rehabilitated and remaining critical areas outside forest land at the end of 1988 fiscal year by region (ha).**

Region	Area planted	Remaining Critical land	
		Directorate Fodder Crops	Directorate Regreening
Sumatra	1,323,003	2,076,400	2,298,600
Jawa	3,045,126	552,500	1,188,500
Nusa Tenggara	468,811	1,163,200	1,225,900
Kalimantan	137,693	731,100*	1,165,300
Sulawesi	835,016	839,200	965,200
Maluku	4,896	305,400	330,400
Irian Jaya	-	62,200	95,800
Total	5,814,545	5,730,000	7,269,700

* Area of Kalimantan Timur adjusted from 13,000 to 413,000 ha.

The records on rehabilitated land during the last five years as kept by the subdirectorate of Regreening and the Directorate of Fodder Crop Development are generally in good agreement. According to these records, the rate of rehabilitation outside forest land was about 160,000-175,000 ha per year. At this rate it would take some 40 years or so to complete the task of rehabilitation, provided no increase in critical land occurs during that time.

The target of rehabilitation during the Fifth Five-Year Development Plan is, according to the Directorate of Fodder Crop Development, about 350,000 ha per year and would reduce the time of completion by half to some 20 years. The target of the Directorate General of Reafforestation and Land Rehabilitation is more ambitious. It intends to rehabilitate 4.9 million ha during the Fifth Five-Year Development Plan, or 980,000 ha per year, a rate which would reduce the time of completion to six to seven, or maybe 10 years. The difference in the target figures does not reflect a difference in scope but more likely a difference in the availability of funds. It will be necessary to provide for the correct distribution of adequate funds to all partners in the rehabilitation work so as to ensure increased and concerted activity in this important field.

Industrial Timber Estates outside Java

The Industrial Timber Estate Programme (Hutan Tanaman Industri, HTI) was initiated in 1984 at the start of the Fourth Five-Year Development Plan with the objective of accelerating the regeneration of secondary forest and degraded areas within forest land outside Java by establishing timber estates to be managed on a sustained-yield basis similar to the timber concessions. According to the technical guidelines, production forest with a stocking of less than 16 m 3/ha, shrub, and suitable grassland are to be converted into plantations.

Residing concession holders, state enterprises, provincial forestry offices, and other entities may establish timber estates. Recent ministerial instructions accord priority to pulp and paper and rayon industries. Three companies are known to invest in pulp and paper plantations in Sumatra and Kalimantan with a fourth one likely to build a factory in Irian Jaya.

Since the initiation of the programme, a total of 68,700 ha of plantations have been established. Of these, some 1,900 ha are rattan that is underplanted in natural forest of suitable plantations. Table 11 shows the extent of industrial timber estate by region.

The preliminary target of the Directorate of Industrial Timber Estates foresees the establishment of 1.5 million ha of plantations during the Fifth Five-Year Development Plan (1989-1993). The target is shown by province and year in Table 12. However, the target for the Fourth Five-Year Development Plan (1984-1988) was also 1.5 million ha but only 68,700 ha (i.e., 66,800 ha without rattan), or 4.5% of the target, was achieved. If it continues at this pace, the 6 million ha by the year 2,000 target of the National Timber Estate Development Plan will never be achieved.

The Draft Forest Plantation Report of the project mentions a number of reasons for the shortfall and cites the more business-oriented ADB Plantation Development Plan that forecasts the establishment of 196,000 ha of plantations within 15 years on six project sites under the direction of Inhutani I, II and III. It proposes a national plantation programme that will reach the target of 6 million ha only some 25 years after the year 2000. According to this programme the extent of plantations by year would be as shown in Table 13.

Table 11. HTI plantations established 1984-1988 by region and species group (ha).

Region	Conifers [*]	Slow hardwoods	Fast hardwoods	Rattan [+]	Total
Sumatra	-	2,762	7,407	-	10,169
Nusa Tenggara	-	975	25	-	1,000
Kalimantan	3,557	15,273	35,741	1,923	56,494
Sulawesi	-	-	1,076	-	1,076
Total	3,557	19,010	44,249	1,923	68,739

[*] Pinus sp. including 7 ha of Agathis sp.
[+] Underplanted in natural forest or plantations.

Table 12. Preliminary targets for HTI plantations during the Fifth Five-Year Development Plan by region (ha).

Province	Fiscal year					Total
	1989	1990	1991	1992	1993	
Sumatra	47,600	102,600	153,500	170,650	170,650	645,000
Nusa Tenggara	4,000	7,850	11,900	13,125	13,125	50,000
Kalimantan	45,000	87,800	132,000	146,650	146,650	558,100
Sulawesi	6,000	19,550	29,300	32,575	32,575	120,000
Maluku	500	17,100	25,600	28,400	28,400	100,000
Irian Jaya	-	5,100	7,700	8,600	8,600	30,000
Total	103,100	240,000	360,000	400,000	400,000	1,503,100

Table 13. Projected forest plantations for different timber utilization (area in ha by year).

Plantations for	2000	2010	2020	2030
Sawlog & veneer and pulpwood thinnings	240,000	1,150,000	3,090,000	6,150,000
Pulpwood only	350,000	760,000	1,170,000	1,770,000
Total	590,000	1,910,000	4,260,000	7,920,000

7. Conclusions

The role of the forest and forestry in present and future economic development is becoming more and more important, both to provide marketable and non-marketed (environmental) goods and services for the benefit of the people. Hence, the acceleration of forest degradation should be overcome by economic as well as noneconomic incentives. In this respect conservation measures, reafforestation, and afforestation should be promoted by inviting the private sector and enhancing the people's participation.

In relation to such efforts, information on the total area of forest plantations in Indonesia is relevant. As of the end of the fiscal year 1988 forest plantations appeared to be some 7.0 million ha (Table 14). This estimate consists of the gross plantation area of Perum Perhutani on Java and the timber estate plantations of the Outer Islands, the rehabilitation plantations within forest land stated in area equivalents of estimated successful plantations, and the planted critical areas outside forest land where the average rate of success (survival rate) is about 50% to 60%. Part of the recorded plantations have undoubtedly been established under the Taungya system, but the so-called buffer zones established through social forestry activities are not included.

An unknown proportion of reafforestation within forest land is for timber production and most of the other plantations for rehabilitation will certainly serve as a source for firewood, fodder and fruits.

It is recommended that forest land use mapping of the National Forest Inventory include an inventory of established forest plantations by their major categories. The inclusion of land use type that may be considered for re- or afforestation has already been mentioned, as has the assessment of critical land. The breakdown of these types of land system, altitude, and slope will provide the first step in a land capability classification for the general planning of reafforestation and regreening activities.

Table 14. **Area of forest plantations by category and regions at end of fiscal year 1988 (ha).**

Region	Timber production *	Within forest land +	Outside forest land #	Total
Sumatra	10,169	341,530	1,323,003	1,674,702
Java	12,364,339	-	3,045,126	4,409,465
Nusa Tenggara	1,000	72,000	468,811	541,811
Kalimantan	56,494	80,850	137,693	275,037
Sulawesi	1,076	200,470	835,016	1,036,562
Maluku	-	420	4,896	5,316
Irian Jaya	-	-	-	-
Total	1,433,078	695,270	5,814,545	7,942,893

* Timber estate (HTI) including 1,923 ha of rattan in Kalimantan; in Java gross area of forest plantations of Perum Perhutani including 8,687 ha of <u>Rizophora</u>.

\+ Area equivalent of successful plantations, part of it for production.

\# Area planted, approximate survival rate 50-60%.

Yield studies and spacing and thinning trials will be necessary for all major tree species in the various regions, altitudes, and sites as the basis for future management of forest plantations. These studies will support in future site productivity surveys that are required for effective extension of the industrial timber estates.

Business considerations and the location of forest industries will determine the siting of future industrial forest plantations. In the process it will be inevitable that natural production forest is also converted to plantations. However, it is necessary to recall the spirit of Ministerial Decree 320/Kpts-II 1986 that states that one of the objectives of establishing industrial forest plantations is to enhance the productivity of hitherto unproductive forest land to produce raw material for wood industries.

It is recommended that areas for reafforestation be selected according to an integrated regional development plan, that the area must be approved by the Forestry Service, and that reafforestation activities be closely monitored. Priority should be given to the rehabilitation of secondary forest, shrub, and grassland with the aim of forming a plantation zone as a buffer between inhabited land and natural production forest.

The deliberate felling and degradation of primary forest for reafforestation purposes must be avoided. The reasons for this practice are often the high costs of establishing plantations in abandoned land where only moderate growth can be expected while clearing of primary forest provides income from timber utilization and promises good plantation performance.

In line with the above proposals, it is suggested that the cooperation between Asian and Pacific countries in exchanging experience and technical know-how on forest and forestry development should be enhanced in the future. In this respect, Japanese expertise and funding capability could be shared with other neighbouring countries, especially in the field of research and training related to the role of forests in environmental protection measures. However, this effort should be designed according to regional and local problems and forest conditions as well as the needs of the respective countries.

Bibliography

Anwar, A. *Economic, social and cultural considerations in sustainable forestry development*. A paper presented at the Seminar on Forest Management to Enhance Sustainable Regional Resource Development of Riau Province, Sumatra, July 26, 1990.

Armitage, I. and Kuswanda, M. *Forest Management for Sustainable Production and Conservation in Indonesia*. Jakarta: FAO; 1989.

Arnold, T.W. *An Analysis of Trade Policies Relating to Indonesia's Forest Products*. Jakarta: FAO; 1990.

Chandrasekharan, C. *Forestry Studies*. Jakarta: FAO; 1990.

Gray, J.A. and Soetrisno, H. *Forest Concessions in Indonesia: Institutional Aspects*. Jakarta: FAO; 1989.

Ministry of Agriculture. *Managing Environment and Natural Resources to Serve Sustainable Agriculture*. Jakarta: Departemen Pertanian; 1992.

Ohlsson, B. *Socio-Economic Aspects of Forestry Development*. Jakarta: FAO; 1990.

Sutter, H. *Forest Resources and Land Use in Indonesia*. Jakarta: FAO; 1989.

World Bank. *Indonesia Forest, Land and Water: Issues in Sustainable Development*. 1989.

COUNTRY PAPER: REPUBLIC OF KOREA

Young-Kyoon Yoon
Assistant Director
Forestry Administration
Seoul
Republic of Korea

1. Forest Environment and Resources

Forest Soil

About 66% of strata in Korea was molded in the Cenozoic era, and granite and gneiss occupy more than 70% of all mother-rock. Due to a changeable continental climate and summertime torrential rain showers, the soil is susceptible to the weathering process and erosion. The majority of forest soil is acidic sandy loam. Generally, shrubs and grasses occupy slopes with shallow soil, while pure pine forests inhabit sterile sites. Mixed forests are found on gentle slopes with relatively deep surface soil.

Forest Distribution

Forests in Korea can be largely divided into warm-temperate, cool-temperate, and frigid-forest zones.

Warm-temperate forest zone: The warm-temperate zone covers the area south of the 35th parallel: a narrow southern coastal region, Cheju Island, and a large number of small southern islands where the annual mean temperature is higher than 14°C. The representative forests in this zone are evergreen broadleaved forests. The majority of natural stands were destroyed by overexploitation and forest fires, however, and subsequently transformed into deciduous broadleaved forests, conifer-broadleaved mixed forests, or pine forests. The characteristics species are *Quercus acuta*, *Castanopsis cuspidata*, *Camellia japonica*, and so on.

Cool-temperate forest zone: The cool-temperate forest zone covers the area located between 35° and 43° north latitude except for the mountainous highlands. The annual mean temperature ranges from 6°C to 13°C. This forest zone is generally divided into three subdivisions: northern, central, and southern temperate forest subzones. The representative forests are deciduous broadleaved forests, but most stands were destroyed and transformed into pine forests. The resultant pine forests have usually developed into the subclimax stage. Further destruction of some forests has provoked problems of soils erosion. Dominant tree species in this zone are *Quercus*, *Betula*, *Zelkova*, *Fraxinus*, *Pinus densiflora*, *P. korainesis*, *P. thunbergii*, and so on.

Frigid-forest zone: The frigid-forest zone, called the coniferous forest zone, covers the northern extremity of Korea and high mountain areas where the annual mean temperature is below 5°C. The representative forests are coniferous. When the forests were disturbed by overexploitation and forest fires, they were usually transformed into deciduous broadleaved forests composed of _Betula_, _Populus_, and so on. Subsequently, some forests gradually changed into larch-broadleaved mixed forest or turned back to pure larch forest.

Reafforestation and Forest Resources

The history of official reafforestation in Korea dates back to the 11th century. However, systemically planned, large-scale reafforestation began in 1973 with the First 10-year Forest Development Plan. Under this master plan, a nationwide tree-planting movement was established and as the result of continuous efforts, the percentage of plantation rose to 31% as of the end of 1990. At that time, forest land in Korea was 6,565,454 ha or 66% of the total land area. However, forest land per capita is no more than 0.15 ha, which is equivalent to only one-fourth of the world mean. Forest land area by ownership shows that national forest land is 21%, and public and private forest is 8% and 71%, respectively. The total stock volume is 248 million cubic metres and the average stock volume per ha is estimated to be 38 cubic metres.

2.   Forestry Policy and Development Plan

The First 10-year Forest Development Plan

A turning point in the history of forestry policy in Korea was witnessed in 1973 with the initiation of that 10-year Forest Development Plan that conceived the rapid reafforestation of the entire country through a national campaign. This development plan was preferentially carried out with the planning of a reafforestation programme for 1 million ha over the country.

The development plan also featured: 1) participation of the entire nation in the reafforestation project; 2) planting of economically valuable tree species to increase village income; and 3) planting of fast-growing tree species. The plan was originally scheduled for 10 years from 1973 to 1982, but was completed in only six years with overwhelming enthusiasm and cooperation between government and the people with the following results.

Establishment of a nationwide tree-planting system: The government established a national tree-planting period from March 21 to April 20, the most suitable period for tree planting in Korea, and promoted tree planting as a nationwide campaign through various units such as villages, corporations, families, schools, and various social groups.

Early revegetation of unstocked areas: As a consequence of nationwide efforts, a total of 1.08 million ha of forest land was planted with trees in only six years. To promote early revegetation of unstocked areas, numbers of fast-growing and nut-bearing trees were planted rather than slow-growing timber tree species in the ratio of 7:3. During this six-year period, most of the denuded forest lands in Korea were successfully afforested.

Establishment of a tending system: In addition to the tree-planting efforts in spring, postplanting care in autumn was encouraged by the declaration of "Tree Tending Day" on the first Saturday of November every year. This nationwide campaign period lasts for one week, during which time various tree tending activities are performed: fertilization with compound fertilizer pellets, weeding, control of diseases and pests, and preparation for the winter. To obtain maximum efficiency from this forest tending work, a responsible postchecking system on plantations is also emphasized.

Table 1.   **Accomplishment of the Forest Development Plan**.

Period	Reforestation	Tending	Erosion Control	Nursery Operation
1st ('73 - '78)	1,080,00 ha	4,177,000 ha	42,000 ha	3,054,000,000 stocks
2nd ('79 - '87)	1,075,000 ha	7,700,000 ha	35,000 ha	1,712,000,000 stocks

The Second 10-year Forest Development Plans

Even though the First 10-year Forest Development Plan resulted in successful reafforestation of the entire forest land in Korea, the quality of the forest was not satisfactory from the economic aspect. Therefore, to maximize forest productivity in the future, the government set up the Second 10-year Forest Development Plan starting in 1979 (Table 1 and Fig. 1). The second plan period was the 10 years from 1979-1988. Its targets were reinforcement of the national reafforestation programme and creation of new economic regions with the establishment of productive forests.

With regard to economic importance and public benefits, all forest lands were classified into reserve forests and semireserve forests based on slope and other environmental restraints. Reserve forests are managed for timber production and public benefit. Therefore, transforming these forests to other purposes is strictly limited. Semireserve forests can be managed for multiple uses such as agroforestry, farm lands, or grasslands, etc.

To establish a new economic forest region, a total of 80 large-scale commercial plantations on 400,000 ha will be established with economic tree species such as Korean white pine, larch, and pitch-loblolly pine.

Erosion control will be undertaken with the principle of complete restoration of eroded areas. Forest research will focus on the development of new native varieties suitable for the Korean soil and climate. Forest protection against humans, diseases, and insect damage will be based on the regulation of human entry into the forests, and the use of natural enemies for insect control. The government support of the Forestry Association Union and Private foresters consists of a 50 billion won forestry development fund during the period (US$1 was about 750 won in 1991).

Recent Forestry Policy

During the latter part of the Second 10-year Forest Development Plan, the government gave priority to the economization of existing forest resources, with emphasis on the establishment of an economic forest sector, multiple use of forest lands, and enhancement of public benefits from forests. These objectives will be accomplished by devoting efforts to sound forest management, control of forest diseases and insect pests, and practical scientific research.

Reafforestation is planned considering the site conditions of the given forest land. Based on the data from a field survey, 21 promising tree species are matched with environmental requirements, such as topography (valley, lower or upper land), soil fertility, and so on. Planting density is generally standardized: 3,000 stocks per ha for slow-growing timber tree species, 400 to 600 stocks for fast-growing tree species, and 400 stocks for nut-bearing tree species. However, there can be some variation in planting density according to the particular tree species and to the objectives of plantation management. The favourable tree-planting season in the central area of Korea is the period of mid-March to mid-April, while tree planting can be started in late February in the southern area.

To reduce the expense of reafforestation, the need for human labour, and the disturbance of forest ecosystems, the tending of natural forests is being promoted. Suitable forests for tending are selected on the basis of the site quality, being converted into productive forests through proper intermediate treatments. To produce high-quality timber, various silvicultural activities are being carried out. To encourage the effective improvement of forest resources, the government provides private forest owners with various financial incentives and technical guidance.

To stabilize devastating torrents, stabilization structures and erosion control dams are built under the principle of complete stabilization of each area. The government intends to prevent disasters completely on devastated forests lands through disaster prevention projects, through which enrichment of the landscape and additional land conservation will be concomitantly obtained.

The major cause of forest fires is manmade accidental fires. Therefore, the government emphasizes a campaign for fire prevention through mass communication and education. Since 1981, the general trend was for decreased forest fire frequency. However, successful reafforestation of forest lands during the First and Second 10-year Development Plan made fires spread faster and wider than ever. Therefore, the government has devoted major efforts to activities to prevent forest fires.

The Third Forest Development Plan

During the 1st and 2nd Forest Development Plans from 1973 to 1987, the spirit of forest-loving and nature protection was inspired through the pannational tree-planting campaign initiated by the government, and rapid forestation of the entire country was completed. Even though revegetation was successfully accomplished, most of the forests are still young. In addition, there are many problems, such as low forest road density, small-scale forest ownership, large proportion of timber import for supply, lack of forest labour, and low income in rural areas, and increasing demand for benefit due to urbanization and industrialization. In order to cope with those problems and to increase the income and public benefit of forests, it is necessary to activate a system of forest management. Therefore, we are required to transform the forest policy from "reafforestation" to "economic forestry."

The objectives of the new plan are to maximize the efficiency of forest land through rational forest land use, to establish the foundation of forest management for a stable supply of forest products, to create income sources in mountain areas, and to promote the public benefit of the forest. The keynote of the policy is to change from a reafforestation policy such as conservation and restriction of forest operations to resource enrichment through management promotion. The period of the plan is from 1988 to 1997 (10 years).

During the period, the foundation of forest resource development will be established. Reserve forests will be managed to enrich forest resources for timber production and public benefit. An intentional land use system for semireserve forests to provide agricultural or industrial land based on effective use of land resources will be established. In order to improve private forest management, forestry promotion areas are designated in private forests, and finance, technology, and tax reduction supports are reinforced. In national and public forests, large-scale forests will be established and the management foundation will be improved and modernized through mechanization and rearrangement of management systems. To increase timber production and pubic benefits of forests, multipurpose management will be adapted.

The emphasis is placed on a stable supply of industrial timber, maintenance of a reasonable timber price, and establishment of a forest product distribution system such as collection, shipment, and joint marketing by forestry associations. In order to acquire continuous and stable timber imports, overseas forest exploitation and export of forest products will be expanded.

Multiincome sources yielding early returns from forestry should be developed to induce villagers to settle in rural areas. Management plans should be set up to develop forestry income sources such as pastures, orchards, special industry for by-products, grazing in forests, and recreational forestry. The government gives financial support to promote those activities.

To promote the land and water conservation function of forests, efficient erosion control, watershed management, and prevention of forest disaster (insect pests, diseases, and forest fires) will be conducted. To provide a pleasant natural environment for a better quality of human life, scenic forests around cities and recreational and educational facilities in forests will be expanded.

Breeding of new, superior tree species using high technology and by setting up techniques for good timber resources will be continued. The efficiency of forest operations by extending forest roads and mechanization will be improved. International competition in the forestry sector will be overcome with rational forest management. The utility of forest products will be increased by utility development projects, and techniques for transformation of forest resources into energy will be developed.

The structure and system of the organization will be adjusted so that the Forests Association is operated by forest owners. The foundation for self-supporting operation will be established through management substantiation and the execution-by-proxy and technical extension service will be activated. To induce investment in forestry, the government will amend the forest laws and improve the forest tax. Increase in the forest development fund and multiplication of loan fund sources will be accomplished by monetary system improvement.

Fig. 1.

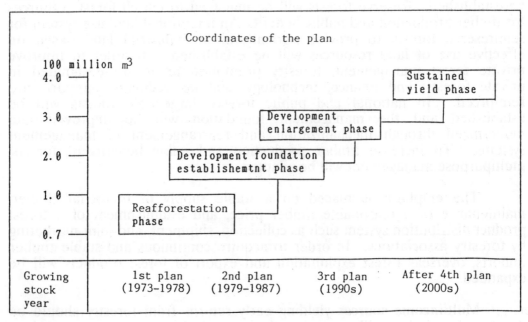

- 182 -

COUNTRY PAPER: PAKISTAN

Mohammad Irfan Kasi
Managing Director
Ziarat Valley
  Development Authority
Government of
  Balochistan Quetta
Quetta, Pakistan

1.   Introduction

Pakistan is a country with a forest deficit. Ten million ha of land in Pakistan, or about 11.5% of the total area, is under the control of government forest departments. However only, only 40% of this (4.9% of the total area) is forested, the remainder being range land. The majority of the forests are coniferous or scrub, accounting for about 40% each, with irrigated plantations, and riverine and coastal (mangrove) forests making up the rest. In addition, considerable tree growth is found on the many farms throughout the country, and there is a tree-planting scheme along roads, canals, and railways primarily for aesthetic use.

Hill forests are found in the moist-temperate, dry-temperate, and subtropical zones and cover 1.78 million ha. These forests are the main source of construction timber, large quantities of fuel wood, and resin. They also sustain the grazing requirements of millions of cattle, goats, sheep, and camels. Increasing human and cattle populations coupled with excessive exploitation are major threats.

Pakistan is a country with a wood deficit. Production from both state and private forests falls short of national requirements, necessitating some imports. The per capita consumption of timber, however, is the lowest in the world. The total annual requirement of timber in the country is about 2 million cubic metres today, by the year 2000 it is likely to be nearer 4 million. Pakistan currently produces just over one million.

Total consumption of firewood in the country is estimated at 16.6 million cubic metres, of which 1.6 is supplied by state forests, but the rest comes from farmlands. At present, firewood accounts for 50% of domestic energy requirements, fossil fuels 13%, and cow dung and agricultural residues 37%. The existing government-controlled forests even at their maximum productivity cannot meet the present requirements of wood and wood products.

In Baluchistan Province of Pakistan, juniper is the single largest block of forest. Juniper trees are considered to be among the oldest in the world. Legend has it that they can live for several thousand years. The forest is presently severely depleted as the following report reveals.

2.   Ziarat Juniper Study

Importance of Juniper Forests

The juniper tree (_Juniperus excelsa_) belongs to an ancient group of plants. There are symptoms of its degeneration in the process of evolution but preservation of this plant has special significance due to its watershed and conservation values.   Internationally and nationally, juniper forests in Baluchistan are very important as they for one of the biggest blocks of the species in the world.   Although this species is also found in the northern parts of Pakistan, its distribution there is confined to comparatively small areas.   In Baluchistan, the significance of juniper forests further increases since it is a forest-deficit province.   It is thus a rare resource.

Juniper trees are xerophytic as they require little water and their transpiration rate is very low.   The role of trees in maximum penetration of precipitation, both snow and rain, is well documented.   Consequently, they play very important roles in groundwater recharge leading to a regular supply of water to support orchards and irrigated agriculture in the juniper tract as well as outside.   Their depletion could be damaging to orchards and the local economy, especially after the accumulated groundwater reserves are exhausted and the rate of fresh recharge slows down.   Their positive effect on macro and micro climates by intercepting clouds for more rain and reducing aridity by maintaining higher relative humidity is commendable.

The direct socioeconomic values of juniper forests to communities include fuel wood energy, grazing for livestock, construction and other timber, fencing material, bark for thatching of huts, medicinal plants for domestic use and sale, shade for humans and livestock, privacy of huts, benefits from the collection and sale of juniper berries, cumin, etc.

Their importance for biodiversity is great due to unique flora and fauna, some of which is specific to the juniper tract.   The area contains such prized species as the Himalayan black bear, markhors, especially the distinctive subspecies of Chiltan markhor and straight-horned markhor, as well as species of birds including the streaked laughing thrush, the Simla black tit, and the bartailed tree creeper.   Its fauna also includes the pallas cat, the marbled polecat, and the Afghan pika as well as Blanford's fox.   Breeding birds include the white-winged grosbeak, the olivaceous leaf warbler, the plain leaf warbler, and the white-cheeked tit.   The endangered Saker falcon and the rare solitary snipe are also important.   The reptilian fauna including snakes, lizards, geckos, and agamas is very interesting.   Among the wild plant species, _Acantholimons_, _Ferula Oopada_, and foxtail lily as well as iris, allium, and tulip species are worth mentioning as they make the landscape spectacular.

The juniper tract provides great opportunities for countryside recreation and tourism as the juniper trees, water points, mountainous landscape, and passes (Tangis) are great attractions.   In addition, the cool climate of the tract provides an opportunity for the urban dwellers of Sind and Baluchistan to escape the scorching heat. If appropriate ecotourism development is launched aimed at benefiting the local populace, some of the benefits of juniper forests will soon become very clear to the people of the tract. Finally, the juniper forests are yet to be explored and used for many scientific and educational purposes.

Special Features of the Juniper Forests and the Tract

The juniper forests have special features that make them a unique ecosystem. The trees survive in an arid climate with an average annual precipitation of 200-400 mm. They occur at high altitude in calcareous and shallow soils in an area where the growth season is short and erosion is active, reducing soil productivity and growth rate. Their preference for soils from the parent material of limestone and shales restricts their distribution. Since conditions for growth and regeneration are difficult, they tend to live a very long time.

The natural regeneration of junipers is not satisfactory. The conditions have been made worse by unwise human interference that is rapidly increasing with the increase in the population of people and livestock in the juniper tract, putting more pressure on the resource. The tribal system helped the juniper forests in the past to survive, but conditions changed when the forests were handed over to the government for creating employment and passing on the responsibility for preservation in the early 1950s. Community as well state forests have also been damaged as a result of weakening of the tribal system. Proper manning of jobs by suitable persons is not possible in the tribal system as the lower jobs are assigned in consultation with Sardars and Maliks and the jobs are inherited.

The juniper forests are of four types as far as their legal status is concerned. The government-owned forests have been declared as protected forests under the Pakistan Forest Act, 1927. The rights for taking fuel wood or timber or grazing of livestock have not been settled. Rather, it has been assumed that the forests are free of rights. However, there is a provision that forest officers can allow removal of wood at their discretion. In practice, the removal of forest produce is not regulated by the Forest Department. In addition to the protected forests, juniper trees occur on a vast tract starting from Koh-e-Siah to the mountains north of Zehri in Kalat District. The Juniper Reserve Trees Act, 1974, prohibits cutting and removal of juniper trees from all areas, irrespective of ownership, without the permission of forest officers but the enforcement mechanism has not been instituted by the Forest Department so far.

The Baluchistan Forest Regulation, 1890, is applicable to the juniper forests in other districts. Government-owned forests are called state forests under the regulation. State forests did not have any right of removal of wood but had limited grazing rights in some forests until 1953-1954 when many state forests, especially in Sibi and Loralai Districts, were opened to local people for removal of wood and unlimited grazing in lieu of their agreeing to declare new areas as state forests. Obviously, the Forest Department gained control of large areas of new forests but, in fact, the conservation status of many previously notified state forests was diluted and weakened. As a result, there is no difference between open areas (community or Guzara forests) and state forests in terms of rights of wood removal and unlimited grazing. Acquiring new forest areas to add to the government estate forfeits the purpose if proper management practices are not carried out and if overexploitation is not checked. The ultimate goal should be the long-term sustenance of forests for communities. There are three types of juniper forests in terms of their legal status: state forests without any right of removal of wood or grazing; state forests open for wood collection and unlimited grazing; and the Guzara forests. The Juniper Reserve Trees Act, 1974, is applicable to the community/Guzara forests but its enforcement has either not been started or is being done half-heartedly. As a result, the Guzara forests are being depleted rapidly. The protection status of the state forests burdened with rights is also the same.

Threats

Overexploitation, unwise land use changes, natural diseases, and absence of natural regeneration in many forest areas threaten the survival of the forests. The juniper forests have been meeting the traditional demands for fuel wood, grazing, timber, and thatching and fencing material but can no longer withstand the increased volumes of removal and overuse. There has been a 3-4% annual increase in the population in the juniper tract. A large migratory population has settled down in orchard valleys and needs fuel wood for heating in winter. The increase in the prices of livestock and their products has provided incentives to the people dependent on livestock to enlarge their herds.

Cultivation or orchard development is good for the local economy but to do it by clearing juniper forests and fencing them by continuous felling of juniper trees is unwise. There has also been expansion in housing as a result of the increase in population and the desire to have better housing, but this activity has consumed large quantities of juniper timber. Construction of roads for developing communications is vital but roads have improved accessibility to forests. Tractors and pick-ups are misused to remove large quantities of wood illegally. A large number of trees have been uprooted in the process of building the roads. At present roads are a constant damaging factor where checkpoints are not established to stop illegal removal of wood.

Juniper trees, especially those under stress, are infested by the dwarf mistletoe, which is plant parasite. As a result, trees growing on large areas in Sasnamanna and Sasnak forests in Ziarat District have died. The juniper is also attacked by heart rat and its berries by a lepidopterous insect. Ninety percent of juniper seeds are not viable. There is also a very low rate of germination of viable seeds, i.e., less than 1% in the best conditions. Then the seedlings are trampled by livestock and poles are cut for fencing, fixing along graves, and for other uses. As a result, the number of seedlings, saplings, and poles surviving in the forests are a very low percentage of the composition of the forests. This is an indication that the juniper forest ecosystem is not healthy and will not survive long to serve local, national, and international needs.

Consequences

The consequences of overexploitation of the resource and of threats mentioned in the foregoing paragraphs are that the forest area is being reduced rapidly, forest cover is becoming more and more open, soil erosion and water runoff are accelerating, soil productivity is decreasing, forests are unable to meet local needs, and conservation, recreation, scientific, and educational values are being lost. Thus, sustainable development in the area will be retarded permanently as the recreation of the ecosystem is either not possible or is very expensive even for a manmade secondary ecosystem.

Although degadation and depletion are rampant, it is useful to identify forests under greater pressure and to launch conservation campaigns there. All community/Guzara forests in all districts of the juniper tract are under great pressure and communities should be helped to ensure their survival and sustainable exploitation. Government intervention for assistance is necessary as these must not be neglected merely for the reason that these forests are not government owned. From among the state and protected forests, the following need immediate attention to halt further depletion:

1) Khushnobe, Bastergi near Ziarat and Chutra, Pil, Kach Mangi, and forests near Manna, Zargi, and Spezandi in Ziarat District;

2) Mazar, Babri, north Zarghoon, and Tagha forests in Quetta District;

3) Khur and Tur forests in Sibi District;

4) Shirin-Kasa, Chutair, Batatair, Bazal Nari, Karbi Kach, and Khummak forests in Loralai District; and

5) Zarkhu and Harboi forests in Kalat District.

Some state/protected forests have been completely depleted. These include Spin Karez, Mari Chak, and greater parts of Zawarkan and Takatu in Quetta District, Narwari in Kalat District, and the western parts of Pil in Ziarat District.

Conservation Approaches

There is need for a change in the attitude of government functionaries as well as communities regarding the ownership and utilization of juniper forests. The forests are for the local communities and are to be managed by them to yield them greater benefits on a sustained basis, irrespective of the present legal status of forests. The communities should be made aware of this change in the attitude of government functionaries and enlisting their cooperation.

It is also important that the objectives of management are well defined and higher priority is given to the intangible benefits such as watershed, recreation, and conservation values in the management of forests. These objectives will supply essential commodities and services such as water for orchards and agricultural crops as well as employment and business in tourism. However, local basic needs such as fuel wood for energy can be met from the forests and where dry wood is no longer available, forests should be closed to wood collection and removal. Letting people take away fire-affected trees encourages them to burn more green trees.

It is essential that alternatives and substitutes are provided to people dependent on dry wood-deficit forests so that the fate of the forests is not further jeopardized and pressure on them is reduced. These could be gas (piped or cylinder), coal, wood imported from upcountry, timber imported from outside, and barbed wire. For timber-made flood protection spurs, wired stone spurs are good substitutes. Popularizing the use of construction material made of iron as a substitute for timber would be useful. Currently, fuel wood and timber are obtained locally from forests at no or comparatively cheap cost. It will be necessary to introduce the element of subsidy for some time, i.e., at least for five years.

The availability of gas cylinders in the wake of a tight supply position of LPG is questionable, but the juniper tract deserves preferential treatment and gas supply could be enhanced either by partly diverting the quota of gas cylinders from other areas, e.g., Quetta City, which have piped gas supplies.

Despite providing alternates and substitutes, the need for local wood and its ready availability will remain. Two workable options are available to enhance local wood resources outside forests by involving farmers and orchard owners. The first option is to plant stream sides, particularly those adjoining fields and orchards. Currently, fields and orchards are prone to erosion during flash floods. Planting of fast-growing species like poplar, mulberry, Russian olive, Robinea, ash, walnut, wild apricot, etc. will not only protect orchards and fields but will also provide wood for different needs of orchards, energy, and timber from trees left longer. Such practice is already prevalent but needs government support to enlarge its scope. The Forest Department must be involved in providing the planting stock, planting, and subsequently irrigating the plants for two summers. Land owners have shown their willingness to protect trees against grazing and help with irrigation when they have surplus water. Where water is not always available, the Forest Department will irrigate the plants with tractor-driven water trolleys filled from dams, karezes, and wells as agreed on by local people. There is perennial water in Khost Nadi and Waluh Nallah in Tor Shore. The Forest Department should not contest ownership of the trees or share in returns; those should belong to the adjoining farmers.

The second option is to provide incentives to land owners to plant forest, fuel wood, and timber trees on their lands. A large number of them, especially the poor and progressive, are willing to participate in tree-planting activities if they are aided in securing additional supplies of water for their fallow fields or new lands. For this, they need government monetary support in water development works such as digging of wells, cleaning or constructing karezes, lifting water from existing wells, building small delay-action dams, and development of springs or construction of tanks. They would agree to plant forest trees on their land at the rate of one acre for each investment of a parcel of 10,000 rupees. They would raise and maintain the trees at their expense for at least 12 years. The Forest Department would not have any share in the produce but would monitor planting and maintenance activities for any violations of the agreement. As the availability and supply of planting material is a key element in successful planting along stream sides or on private land, raising suitable planting material in forest nurseries within the juniper tract is of prime importance. To achieve this objective, the nursery at Quetta could be expanded with funds to be provided under the proposed Ziarat Juniper Project.

Forestry research has so far been confined to making a breakthrough in the artificial raising of juniper seedlings and their planting under very controlled conditions and on a very small scale. Fortunately, the efforts of the Pakistan Forest Institute, Peshawar, have succeeded but now trials at the field and demonstration levels are needed to prove their success in field conditions prevailing in the tract. Secondly, research on supporting natural regeneration of junipers (by seed, coppice, and layering) has not been undertaken so far. Research on many aspects of watershed, range, and wild life necessary for management of the resources needs to be pursued. There is also a strong need for establishing a permanent meteorological station at Ziarat.

The change in social attitudes to inculcate interest in and love of the juniper is necessary to halt further depletion. For this purpose, and awareness campaign for students and the public through such approaches as charts and brochures in Urdu and radio programmes and slide shows in local languages is essential. It is also necessary to have a full-time conservation education officer to develop publicity materials and manage the campaign.

Community participation in the management of forests, whether Guzara, protected, or state forest, is essential and needs to be ensured through local forest committees. These need support from the Forest Department for achieving their goals effectively. The Ziarat District Forest Preservation Committee is a good start but needs to meet regularly and all issues/approaches concerning the forest must be in its ambit. It also needs to be expanded to include suitable persons from all parts of Ziarat District. Establishing such committees in other districts would be very useful.

The present infrastructure is very weak for protecting and managing state of protected forests. It is not geared to helping communities in social forestry or for managing Guzara forests. The staff lack essential facilities of housing, transport, uniforms, equipment, and reasonable travelling allowances. They need in-service training and orientation to discharge their duties. Check points that are important for checking on illegal removal of wood from forests, especially that going outside the juniper tract, need to be established. Thus strengthening of infrastructure has been conceived as an important component of the proposed project. Parts of the juniper forests in Ziarat District are worth declaring as national parks as well as world heritage sites under the World Heritage Convention due to the unique ecosystem and interesting associated species of flora and fauna.

Haphazard growth of Ziarat town is taking place. The growth will accelerate when the constrains on water supply to the town ends in the near future as a result of new water source from tubewells. There are already serious problems of garbage disposal and traffic congestion during the summer season. Expansion of the town is currently taking place at the expense of juniper state forests that are the attraction of the town. The destruction of this resource base defeat the objectives of development. Demarcation of forest boundaries and restricting town growth outside the state forest boundaries are needed. The development of the town should be such that it promotes the local economy, it becomes a tourist town, and outsiders are prohibited from acquiring plots.

Protection Measures Needed

The protection and management of forests will improve with the strengthening of infrastructure and in-service training or orientation of staff. The Forest Department will, however, need external assistance, guidance, and monitoring on a continuous basis for:

1) interaction with farmers and land owners for tree planting along streams and on private land after developing water as it needs an extension service and extra technical skills approach to make it a success;

2) launching awareness campaigns for the public and students;

3) establishing the forestry, watershed, range, and wildlife research capability of the Baluchistan Forest Department;

4) supporting local and district forest committees;

5) designing and conducting appropriate in-service training or orientation courses for staff;

6) arranging for appropriate short foreign training courses or study tours for professionals;

7) arranging and supervising short-term consultancies for forestry, rural development, town planning, etc.;

8) raising funds, especially grants from international agencies for activities during the implementation of the project; and

9) arranging the supply of alternates and substitutes at subsidized rates.

COUNTRY PAPER: PHILIPPINES (1)

Joseph M. Alabanza
Regional Executive Director
National Economic and
Development Authority
Cordillera Administrative
Region (NEDA-CAR)
Baguio City, Philippines

1.   Introduction

This paper draws extensively from two basic documents, namely, the Philippine Strategy for Sustainable Development and the feasibility study for the project entitled _Community-Based Environment and Natural Resource Management Project_ which is currently under evaluation by our agency. This project basically addresses the problems of upland poverty and environmental degradation using the community as a mechanism to institutionalize forest resource management. Adhering to the principles of the integrated development approach, the project has adopted five development components, namely: on-farm development; off-farm development; livelihood component; infrastructure development; and institutional development.

The on-farm component seeks to promote farming systems that provide for stable, sedentary upland agriculture through soil conservation, improved annual cropping/cropping practices, and conversion to perennial cropping or agroforestry on lands not suited to annual crops. Meanwhile, the off-farm component focuses on improved management of the community's natural resource base with an emphasis on the protection and/or management of primary or residual forest and the conservation of upland soils to assure their future productivity. The livelihood component introduces support activities to rural industry development as well as livestock and poultry development as integrated activities in agroforestry development activities. Recognizing the impetus provided by infrastructure to any development activity, an infrastructure component was established specifically to lend support to the activities undertaken under agroforestry, reafforestation, upland agriculture, poultry development, rural industry, and other livelihood activities. Finally, the institutional development aspect draws in detail the project's strategy even as it endeavours to involve the community in the general process of planning, implementing, and monitoring of forest management activities. This paper touches more on the institutional aspect of the project.

This paper is composed of four sections. Section 2 discusses the Philippine environment, section 3 the state of the country's ecosystem, section 4 the Philippine strategy for sustainable development, and finally section 5 the Community-Based Environment and Natural Resources Management Project.

2.   The Philippine Environment

The Philippine environment is classified into six ecosystems, namely: its forests, marginal lands, croplands, coastal areas, freshwater ecosystems, and urban areas (**Fig. 1**). Forests and marginal lands are normally termed the uplands. They comprise 50% of the country's land area. The country's forest cover consisted of 10.823 million ha in 1981. By 1988, this was reduced by 71.7% to 2.911 million ha. This remaining forest cover is about 25% of the total land areas; in the 1950s the cover was about 75%. Marginal lands include denuded forests, grasslands, brushlands, and barren areas. They total 12 million ha. This is about 41% of total land area. The croplands ecosystems covers 34% and is made up of coconut plantations, arable lands planted to cereals and sugar, other crop plantations, and fishponds. The coastal ecosystems consists of 0.6056 million ha and is composed of mangroves, marshy areas, and coral reefs. Rivers and lakes make up the country's freshwater system. There are about 0.3872 million ha under this system. Finally, the urban ecosystem are built-up areas totalling 130,000 ha. Some 20 million people reside therein. By the year 2000, these areas are expected to be populated by 30 million.

3.   The State of the Philippine Ecosystem

We have identified seven problems besetting the Philippine environment. These are: 1) deforestation; 2) migration to the uplands; 3) increasingly heavy use of chemical fertilizers and pesticides; 4) rapid conversion and destruction of virgin mangrove swamps; 5) destruction of corals; 6) pollution; and 7) garbage collection.

Deforestation occurs on 700,000 ha annually and is caused by the following: swidden farming, pests and diseases, forest fires, and logging. These have so far induced erosion estimated to be about 100,000 ha at a depth of 1 m or about 1 billion cubic metres of material annually. Further, this has destroyed the natural habitat of certain indigenous species of bird and animal forms. The loss of this biodiversity has translated into 18 endangered wildlife species. Another 25 are on the threatened list.

The migration to uplands has exerted pressure on the already fragile ecosystem of the uplands. As of 1988, 18 million inhabited the uplands, 8.5 million of whom are within the forested areas and growing by 2.55% annually. This movement/habitation has seen the exercise of agricultural practices not suitable to the uplands as well as tenurial problems in public domain. The latter has further aggravated the problem of encroachment into ancestral lands.

In an effort to spur and increase production, chemical fertilizers and pesticides have been used extensively. Little or no efforts have been undertaken to counter or mitigate their deleterious effects.

While the country boasted of 450,000 ha of virgin mangrove swamps in the 1920s, to date this has declined to 149,000 ha, or a reduction of 67%. These are mostly secondary growth, containing species other than the original mangrove species. Satellite images have indicated 284 square kilometres of sedimentation patterns, indicative of the gravity of the erosion, siltation, and sedimentation in coastal ares. Thirty percent of the country's reefs have been destroyed while the remaining 70% have been subjected to major environmental damage due to gathering, siltation, and destructive fishing. Moreover, these areas are the final destination of most of the mine tailings generated by the mining industry.

Seventy percent of land-based pollution is from domestic sources while 30% is from industrial activities. Apart from these, pollution from ships and oil spills is common. The extent and gravity of air pollution is not well known as the last monitoring activity was in 1983. Estimates, however, show that the air pollution load in Metro Manila comes from vehicle emissions (60%) and industry (40%). The water pollution problem is emphasized by the fact that no city in the Philippines is served by a sewerage system. It is generally believed that all rivers of Metro Manila are biologically dead. The same is true in the surrounding provinces that are believed to be in bad condition. Only ¼ of industrial firms nationwide comply with water pollution control requirements. Salt water intrusion in groundwater supplies is increasing especially in Metro Manila, Cebu and Negros.

The collection and proper disposal of Metro Manila's 4,000 million tons (MT)of solid waste generated daily remains a problem. Not all is collected and some finds its way into the river system. That collected is thrown into open dumps, posing health hazards to the surrounding communities and to those who make a living out of dumped waste.

4.   The Philippine Strategy for Sustainable Development (PSSD)

What is the government's action toward solving these problems? The Philippine government is explicit in its development thrust as it endeavours to move toward: sustainable development of the country's natural resources; access and benefit-sharing in resource use; and an integrated approach to environmental management and protection. Thus far, the most laudable institutional mechanism the government has evolved is the Philippine strategy for sustainable development (PSSD) that projects a comprehensive 10-year blueprint for addressing current environmental concerns. It is a document specifically citing a number of general strategies toward sustainable development. The general strategies expected to be further translated into more specific sectoral strategies are: 1) integration of environmental consideration in decision-making; 2) proper pricing of natural resources; 3) property rights reform; 4) establishment of an integrated protected areas system; 5) rehabilitation of degraded ecosystems; 6) strengthening of residuals management in industry; 7) integration of population concerns and social welfare in development planning; 8) inducing growth in rural areas; 9) promotion of environmental education; and 10) strengthening of citizen participation and constituency building.

The implementation of the PSSD will be a long-term process and many specific actions will entail further study, public debate, congressional decision, and/or prior institutional strengthening. Alternative actions can be taken in the pursuit of the same ends, and the social, private, and budgetary costs involved will limit the choice of actions and the pace of change. Within these constraints, many choices have already been taken to deal frontally with the main issues, and further steps during the two years of 1991-1992 have been identified.

5. The Community-based Environment and Natural Resources Management Project (CBENRMP)

Referred to as the community-based environment and natural resource management project (CBENRMP), the project basically draws from the fundamental philosophy of drawing strong social and institutional development support for effective and sustainable management of the environment and natural resources. This consequently identifies the empowerment of communities and the devolution of power and functions to local government units as critical components in the effective management of watersheds.

Through the Department of Environment and Natural Resources (DENR), the Philippine government was able to negotiate with the World Bank a Project Development Assistance for institutional strengthening and regional community-based development projects. That assistance will support a hybrid operation consisting of a quick disbursing sector adjustment component based on policy and institutional strengthening and a regional community-based resource component activities (**Fig. 2**). It has three components, namely: 1) the integrated protected area system design studies; 2) monitoring and enforcement of forestry laws and regulations; and 3) regional community based resource management component or the community based environment and natural resources management project (CBENRMP).

When we talk of community-based projects, we mean: that events such as planning, project designing, or implementation are participatory, site specific, and localized, i.e., while support for projects may be sourced externality, the control of project processes is internal to the community; that community-based investment projects presuppose a substantial degree of community organizing as a preliminary activity in project institution; and decentralization, in terms of project decisions being entrusted or limited to local communities while project support is entrusted to line agencies.

The CBENRMP will be undertaken in Regions I (northern Luzon), II (Cagayan Valley region), IX (western Mindanao), and X (northern Mindanao) (**Fig. 3**). These regions account for about 43% of the primary and residual forested area remaining in the Philippines. Note that the thrust is on upland management. It rests on the basic premises that: 1) poverty in the uplands has caused deforestation, and that poverty is due to the lack of livelihood opportunities therein; 2) the community that draws sustenance and livelihood from the resource, which in this case is the watershed area, would be in the best position to protect it; 3) environmentally destructive activities in the uplands are difficult to arrest unless they can be replaced with livelihood activities less destructive and at least equally productive; 4) establishment of community facilities and provision of tenure rights also help stabilize communities and increase individuals' time horizons and these mechanisms can therefore constitute incentives for participation in constructive natural resource management activities; 5) participatory uplands resource management recognizes that the ultimate decision makers are the farmers and not the policy makers or planners who hand down a development programme; 6) there is a need to concentrate on building the capacity of local governments and community institutions to sustain, expand, and replicate these efforts as well as for the use of economic approaches to community development and technological extension.

Two basic problems are expected to be addressed, namely: 1) natural resource degradation; and 2) socioeconomic and cultural problems. Within the first problem, the issues to be addressed are forest denudation/depletion, soil erosion, and siltation and pollution of rivers. Within the second, the issues are the decline in agricultural land and farm size, ethnic conflicts, and the physical inaccessibility of the regions. While the scope of the study focuses on both the Ilocos Region and the Cordillera Administrative Region (CAR) as Region I, the present paper refers mostly to the latter.

The Cordillera Administrative Region and its Problems

Three major islands make up the Philippines. However, the ease of administering political affairs gave way to the creation of 14 administrative regions (**Fig. 4**). Thus, within the major island group of Luzon are seven regions (**Fig. 5**) of which the CAR is a part. The CAR is located at the northernmost portion of the Philippines, occupying 1.93 million ha or only 6.1% of the total land area of the country (**Fig. 6**). It has a mountainous topography from which a number of major rivers in northern Luzon originate (**Fig. 7**). In 1990, the regional population totalled 1.2 million, bringing the population density to 64 persons per square kilometre which is the least in the Philippines. Population grew 2.3% between 1980 to 1990. The Philippine population grew at the same rate. The upland population in the CAR grew at a rapid rate in recent years due not only to the rate of livebirths but on account of migration from the lowlands. As of 1989, there were 250,870 forest dwellers, 46% of whom were tillers/farmers in the forests. The CAR's forestland totals 1.4 million ha, 4.9% of this is inhabited by upland population.

While the CAR counts among the regions with the smallest area devoted to alienable and disposable purposes (**Fig. 8**), it has the largest area needing vegetative cover rehabilitation (**Fig. 9**). Extensive deforestation has taken its toll on the productivity of agricultural activities and on settlements

(**Fig. 10**). Soil erosion is widespread. Moderate to severe erosion affects about 67% of the total land area of the region (**Fig. 11**). This is as far as pre-1990 earthquake conditions are concerned. As post-quake data on erosion are not yet available, we might be able to assume that areas then classified to be moderately eroded have degraded and added up to areas already severely eroded. Widespread misuse of inorganic fertilizers, pesticides, and insecticides has threatened human lives either in the form of residues in crops or as runoff in rivers. Other major inorganic pollutants are those embedded in mine tailings discharged into rivers.

CAR's river systems basically serve as agents in transporting sediments, considering that it cradles the headwaters flowing into the river systems of the lowlands (**Fig. 12**). It boasts of nine major river basins from where major rivers, streams, and creeks are drained. Among the significant basins are Abulog-Apayao, Chico, Abra, Magat, and Agno River basins.

Tenancy is low in the region due to the more equitable traditional land use system developed over time to cope with biophysical and social constraints (as usufruct rights to swidden land, traditional exchange labour, borrowing of rice fields from those who have more, with some kind of sharing arrangement or for free). Thus far, limited income can be derived from the cultivation of small parcels of land. A general outlook on the tenure situation in the Cordillera further shows that public land is generally treated as open access, thus making encroachment easily undertaken not only by landless farmers but by established farmers who have noted opportunities in the commercial expansion of vegetable farming. Thus, tremendous pressure to cultivate the remaining hectarage in public lands exists unless alternate opportunities for nonfarm income are introduced.

Ethnic conflicts stem from two factors, ethnic diversity; and traditional lowland-upland animosity over political dominance and control of land and resources. Cultural diversity in the Cordillera is high especially as there are 11 major ethnolinguistic groups, causing competition and conflict over land. This situation has, on the other land, created arrangements to reduce interethnic tensions in the region.

In relation to the animosity between lowlanders and uplanders, it stems from the general perception that the former is viewed as a constituency of the dominant political centre. Having associated lowlanders with the political centre, uplanders generally perceive themselves to be on the periphery of the political centre so that incursions of lowlanders are seen to be a threat over their control of upland resources.

Finally, the physical inaccessibility of CAR is more apparent in this region, due largely to the innate steepness of its topography. This has resulted in marked inadequacy, if not absence, of basic social services. Government presence is perceived as confined to health and education only.

CBENRMP Development Principles

It should be noted that the ultimate objective of the CBENRMP is a sustainable community-based and ecologically sound management of a natural resource base. By the end of the project's lifetime, the following should have either been attained, produced, or established: a sound ecological framework for sustainable biophysical, socioeconomic, and institutional interventions; agroforestry farms and communal forests; self-reliant and environmentally management-oriented community organizations; reduced poverty levels; improved capacity of field agencies of the national government and local government units to support community-based resource management; actively involved nongovernmental organizations (NGOs) in institution-building and development of people's enterprises (**Fig. 13**). These being the objectives, the following development principles were established, after a rapid rural system appraisal was conducted:

1) That the project will improve the utilization and management of natural resources. Its framework rests on assisting selected communities to develop the processes and capability to utilize and manage local natural resources on a sustainable basis.

2) That the project will improve the efficiency of access to public facilities and social services. Of vital consideration is the kinds of public infrastructure facilities and social services and how these relate to projects for agricultural production, agroforestry, reafforestation, and forest management. Also to be considered are basic or elemental improvements in some facilities (such as cleaning irrigation ditches, improvement of farm-to-market roads, or installation of pipes in springs to bring water by gravity to households) which the local community may be able to carry out alone or with government support.

3) That the project stimulates community participation and private-sector and off-farm employment and income. Networks of CBENRMP projects may best be implemented via the project offices approach due to its simplicity and the ease with which to dissolve it. It has become necessary to identify precisely those groups that have failed or are failing to participate in the regional development process in order to elicit community participation. It is also imperative to identify strategies to formulate specific responses designed to contribute toward greater involvement of groups that are slow in participation in local development efforts. Early involvement of the private sector will likely facilitate the implementation of economically promising projects. Further, intensification of sources of off-farm income and employment will help minimize further encroachment into upland resources by drawing them toward off-farm employment opportunities that the private sector can help establish.

4) That the project will build up local institutional and technical capabilities. It is vital that the chosen coordinating institution,

whether at the local, municipal, or provincial/district level, can be as responsive as possible to the distinctive and varied needs of the identified CBENRMP area. Moreover, efforts to involve the constituent population or NGO groups in the design of ENR and support project must be strengthened with a programme to train and equip participants to discharge their responsibilities effectively. This entails the district or provincial coordinating entity to shift emphasis from supervising detailed programme operations to the setting up of policies, staff and community organization, inspection and auditing, among others, or CBENRMP projects. Apart from sufficient funds, the direct local implementer should have some operating discretion in planning and implementing projects. Community organizing may be pursued in terms of identified specialized groups of NGOs to participate in the validation and final identification and preparation of CBENRMP projects. This is crucial for mobilizing local popular support for projects and to conduct projects in ways that reflect their priorities and capabilities. Training of government frontline technicians is also necessary. They must also be equipped, guided, and properly motivated by higher salaries. Responding to the new powers and autonomy granted by the LGC, CBENRMP projects should, as part of their objectives, develop the capacity of local officials to plan, implement, and manage ENR and related projects.

Generation of Area-specific Projects

The CBENRMP followed four stages in the project formulation process. The general process adopted by the proponents in evolving the projects from the time the CBENRMP commenced is depicted in **Fig. 14**.

Selection of Project/programme area: It used the watershed unit as the planning unit and utilized selected parameters for analysis of natural conditions as topography, vegetative cover, and climate as such bases for selection and delineation. The selection of the project areas followed three sieving steps (**Fig. 15**). The criteria utilized in the selection and sieving process are presented in Appendix A. The process involved the following:

1) *Selection of watershed zone.* Out of the 13 zones identified, eight from CAR were considered. These zones were considered to be mostly "critical" and with high-value infrastructure like dams, and are economically depressed.

2) *Selection of municipality or municipalities.* Seven of the 76 municipalities of the region were identified in this process.

3) *Selection of specific sites.* Ten barangays were chosen as the final and pilot sites of the project.

Definition of problems and opportunities: A rapid rural system appraisal (RRSA) was employed to provide the basis for identifying major problems and opportunities related to the CBENRMP in each site.

Project identification, selection, and validation: Identified projects evolved from analysis of problems and opportunities; a large part were mere ideas and need extensive evaluation and validation with community members; some may come across as already ongoing or previously approved for implementation on site; some may have been previously prepared by government officials but like project ideas, need to go through the process of initial preparation and evaluation. Consultations were also conducted with local officials and representatives of community organizations and line implementing agencies to avert conflict and foment cooperation and complementation of sectoral efforts. The process evolved additional criteria which are contained in Appendix B.

Preparation of action plant/project packages: This entailed the submission of the packages to screening, validation, and providing details of prefeasibility study of feasibility study levels.

Target Beneficiaries

The target beneficiaries of the project are: the upland farming households found in the watershed areas, whether occupying and/or tilling public or privately held lands; and the upland community that broadly means the barangay and including relevant sectors and groups such as barangay councils, women, and youth.

Development of Institutions under the CBENRMP

Institutional development activities are but one of the components of the CBENRMP (**Fig. 13**). For a more detailed account of the other components of the CBENRMP please refer to Appendix C. The points offered by the project are of two types: at community level; and at for the overall project management level of the CBENRMP. Thus, at the community level, the points include: utilizing community organizing as a mechanism for institution building. It recognizes that this will empower the community to identify better options for themselves and to organize themselves to achieve political influence and socioeconomic gains. This basically entails the social preparation of the community to become wise users and effective managers of the watershed and at the same time pursue sustainable activities to attain higher productivity and improve socioeconomic welfare. Specifically, this activity includes:

1) broadening the knowledge base of the community on environmental concerns;

2) strengthening its organizational capabilities to pursue watershed management and livelihood activities; and

3) developing and enhancing technical skills in agriculture, agroforestry, rural industry, and other livelihood activities and in operating and maintaining support infrastructure works.

Community Organizing Work

First, there is a identification of entry points, in the form of institutions or structures which could be the barangay council and/or the council of elders, existing integrated social forestry associations, or other farmer's/people's organizations. Second comes the identification of potential participants in the CBENRMP project. This entails an inventory of target beneficiaries and the organization/reorganization of target beneficiaries/groups. Major activities pertinent to the latter include the presentation of organizational and legal framework and basic leadership and membership training. Third is the planning for the implementation of accepted project activity. This involves drawing the participation of target beneficiaries and barangays as represented by leadership. This leadership provides legitimacy to the programme and is involved in the project implementation phase. Fourth is the conduct of continuous training of beneficiaries in the following areas:

1) organizational strengthening that includes leadership and membership training, financial management, conflict management, cooperative development, networking and linkaging, and fund sourcing; and

2) technical skills training in agriculture, forestry, environmental management and protection, other livelihood projects, and entrepreneurial skills including project identification and development, accessing capital support, and marketing.

Training activities are not confined entirely to the target beneficiaries as the project endeavours to upgrade the skills of the involved government organizations as the project's implementers.

Sixth, is the setting up of a mechanism to allow participants to monitor and evaluate the implementation of project activities by themselves. As much as possible, the focus is on the generation of information that will determine the overall impact of the project on the beneficiary as an individual and on the community as a whole.

At the outset, the CBENRMP takes cognizance of the critical role played by NGOs in community organizing. In general, it is assumed to pursue the task of community organizing and strengthening the organizational capabilities of the community groups.

CBENRMP Project Management

At the overall project management level, the CBENRMP adheres to the management principles of: participatory or grassroots planning and implementation; imparting a process and learning perspective; recognition of the support role of government, NGOs, and academic institutions; and decentralization and devolution. The institutional set-up established for the management of the CBENRMP is summarized in **Fig. 16**.

6. <u>Conclusions</u>

The underlying philosophy of environmental protection and poverty alleviation guiding the CBENRMP is sound, especially as it seeks to address two of the fundamental problems besetting the country. The focus on the people and the community as the user, manager, and ultimately the beneficiary of the gains from effective and sustainable watershed management underscores the project's emphasis on social equity which is an avowed national goal. Correspondingly, the adoption of a participatory approach that emphasizes community organizing in order to equip communities and make them socially and economically self-reliant is consistent with the ultimate goal of social development.

Bibliography

Department of Environment and Natural Resources-Cordillera Administrative Region. <u>Regional Master Plan for Forestry Development</u>, volume 1. Baguio City, 30 November 1991.

National Economic and Development Authority-Cordillera Administrative Region. <u>Regional Physical Framework Plan</u>. Baguio City, 1991.

Orient Integrated Development Consultants, Inc. and Department of Environment and Natural Resources. <u>Region 1/CAR Environment and Natural Resources Management Feasibility Study</u>, Main Report (Final Draft), volume 1, December 1991.

Orient Integrated Development Consultants, Inc. <u>Substantive Proceedings of the First Environmental and Natural Resources Sector Adjustment Program. Launching and National Consultative Workshop</u>. Asian Institute of Tourism, Diliman, Quezon City, 11-12 July 1991.

"Philippine Strategy for Sustainable Development," <u>Cordillera Gangza</u>. Volume 11, No. 2 (April-December 1990): 3-8.

Appendix 1. Criteria used in the selection of priority project sites.

Identification of watershed zones:

1)   broad mountain zones from which water flows as headwaters or tributaries;

2)   erosion as indicated by latest soil maps as high to severe;

3)   zone should include territories of major ethnic groups;

4)   population density is high compared to the regional or provincial average; and

5)   poverty levels are generally high.

Identification of municipalities:

1)   municipality is within the watershed zone;

2)   it is accessible by motor vehicles;

3)   average household incomes are among the lowest in the province;

4)   there have been no reported armed incidents in the past 2-3 months;

5)   residents are receptive to projects;

6)   it has high potential for agroforestry type projects;

7)   no similar efforts are being implemented, especially in agroforestry or forestland agriculture, and if there are, they are supportive of livelihood-cum-watershed improvement project types;

8)   there are active organizations (government organizations, NGOs, and private organizations);

9)   it displays prototype watershed and poverty problems in the zone;

10)   leaders are likely to support and project; and

11)   it counts among the priority area identified in the regional or provincial development plans.

Identification of specific sites (barangay level):

1)   there exists substantial uplands;

2)   upland areas are highly to severely degraded watersheds;

3)   streams flow from the site going toward previously identified river systems;

4)   infrastructure indicators (i.e., small and temporary houses, poor road systems, no large stores) suggest a high incidence of poverty;

5)   site is acceptable to local field agencies of the national government; and

6)   receptiveness to government projects undertaken on site is evident.

Appendix 2. Additional criteria utilized in project evaluation via the consultation process.

1) The project is small-scale with a high potential for replication.

2) The project is highly demanded and appreciated by local people.

3) The project will benefit low-income groups/families.

4) The project has high employment-generating capacity.

5) The project stimulates population deconcentration through the creation of jobs and service facilities outside watershed zones.

6) The project utilizes relatively technologies adaptable to local labour and entrepreneurial capabilities.

Appendix 3. **CBENRMP development directions.**

Problems identified	Objectives	Strategy
I. Natural Resource Degradation a) Forest Denudation/Depletion While the two regions have the smallest area devoted to A & D, both also have the largest area needing vegetative cover rehabilitation. Extensive deforestation on the productivity of agricultural activities and on settlements is high. Among causes of deforestation are: extensive use of firewood for tobacco drying in Region I and open pit mining and the conversion of forest lands to vegetable production in CAR.	Sustainable Community-based and Ecologically Sound Management of a Natural Resource Base Specific Objectives: a) a sound ecological framework for sustainable biophysical, socioeconomic, and institutional interventions; b) established agroforestry farms and community forests; c) self-reliant and environmental management-oriented community organizations; d) reduced upland poverty;	General strategy applies the buffer zone concept whereby a buffer zone is established between the remaining primary and/or vigorous residual forests and areas of multiple use and human settlements. Buffer zone consists of graduated circles of protection around forest areas which generally corresponds to increasing slopes and ecological vulnerability. Site development includes: **GO-NGO Participation.** Go to provide technical development, promotion, and training of extension workers; NGO for community organizing, credit transfers, capital build-up,

b) Soil Erosion

c) Siltation and Pollution of Rivers

Widespread misuse of inorganic fertilizers, pesticides, and insecticides has resulted in threats to humans as either residues in crops or as runoffs in rivers.

Other major inorganic pollutants are those embedded in mine tailings discharged into rivers.

II Socioeconomic and cultural problems

a) Declining agricultural land and farm size.

Limited income can be derived from the cultivation of small parcels of land.

Tremendous pressure to

e) improved capacity of field agencies of the national government and LGUs to support community-based resource management; and

f) active NGO involvement in institution building and development of people's enterprises.

and training of private organizations (POs) in managing income generating projects; POs for development/ management of cooperative business enterprises.

Target Area and Beneficiaries. In the identification of project roles and communities, the most critical step is the prioritization of household categories to determine the frontline participants in the project components.

Land Allocation for Agroenvironmental Projects. Modal upland watershed provides the central basis for the formulation of land use/ activity allocation in the project. Parameters that will dictate the types of economic activities a farm family can sustain overtime will include existing land uses and slope.

Tenurial situation will also influence the final land use pattern. As a result, two major modules were identified: agroforestry modules (on-farm); and community-based modules (off-farm). **Figure 17** summarizes this model.

cultivate remaining lands in the public domain exists (regardless of physical constraints) unless other nonfarm income-generating opportunities are introduced.

b) Ethnic Conflicts

Springs from ethnic diversity and traditional lowland-upland animosity over political dominance and control of land resources.

c) Physical Inaccessibility

Lack of access roads, which may be attributed to steep topography, in highland agricultural areas and settlements more apparent in CAR.

Note: Appendix 3 gives the details of the project components.

Appendix 4. CBENRMP project components.

Off-farm development projects

1) **Roadside afforestation.** Includes the planting of fast-growing fuel wood species along the major access roads of project sites where the demand for the products of these is high but the land area for development is limited. Apart from economic use, roadsides will be stabilized and protected from accelerated soil erosion and landslides.

2) **Riverbank Protection.** Establishes bamboo groves along major river systems in project sites for bamboo pole production and the enhancement of the surrounding environment.

3) **Community-based Rattan Plantation Development.** Establishes rattan plantations to increase the livelihood opportunities of the local community and maximize the ecological benefits from existing secondary forests and/or established forest plantations through the enrichment of vegetation.

4) **Community-based Agroforestation.** Establishes agroforestry crops on presently unoccupied, uncultivated, and/or abandoned areas where less important vegetation species like grasses grow.

5) **Community-based Reafforestation.** Rehabilitates degraded watershed sites critically located on very steep slopes and which pose environmental risks to framing activities and settlement areas.

6) **Community-based Forest Resource Management.** Assumption of responsibility and control over the resources by the local community within the project site.

On-farm development

1) **Training.** This is to be undertaken in established agricultural areas where farmer-beneficiaries and other farm communities will be provided training in basic farming skills relevant to existing crops.

2) **Farm Demonstration and Research on Crop Rehabilitation and Farm Productivity Enhancement.** These are to complement the training programs established in (1).

3) **Integrated Agroforestry-cum Livestock.** This is to be undertaken in areas with slopes higher than 18% but less than 50%.

4) **Food-Fuelwood-Livestock Development Project.** To be undertaken on underutilized lands within the 8-10% slopes. This project provides the central strategy for the mitigation measures for the expansion of economic activities outside A and D lands.

Other livelihood projects

LIVESTOCK AND POULTRY DEVELOPMENT. Involves the integration of intensive livestock production with agroforestry development from which livestock subsist on fodder and multipurpose tree legumes and improved grasses used as hedgerows and crop residues from alley cash crops.

On-Farm Upland Agricultural Development.

* Integrated Agroforestry Farm-Development cum Livestock (IAFL) - integration of cow-calf operation and draft carabao with agroforestry farm development.

* Integrated Small Livestock and Community-based Vegetable Production and Processing - integration of small livestock such as swine and goats with intensive vegetable production.

. Off-farm Support Resource Development.

* Credit Support

. Training

. R & D

RURAL INDUSTRY
Involves the provision of assistance in the following forms:

. Skills Development (design/processing) and Entrepreneurship (business management) Training.

. Extension/Technical Assistance for Product Development, Market Linkaging, and Price Monitoring.

. Credit Support and Sourcing.

Infrastructure development

These are activities that will lend support to the major components of agroforestry, reafforestation, upland agricultures, livestock development, rural industry, and other livelihood projects.

1) Improvement/Upgrading of Existing Road.

2) Construction of footbridges.

3) Improvement/Construction of Access Graded Trails.

4) River Bank Revetment.

5) Construction of Small Irrigation Systems.

6) Construction of Domestic Water Supply Facilities.

Fig. 1. Typical Philippine environment.

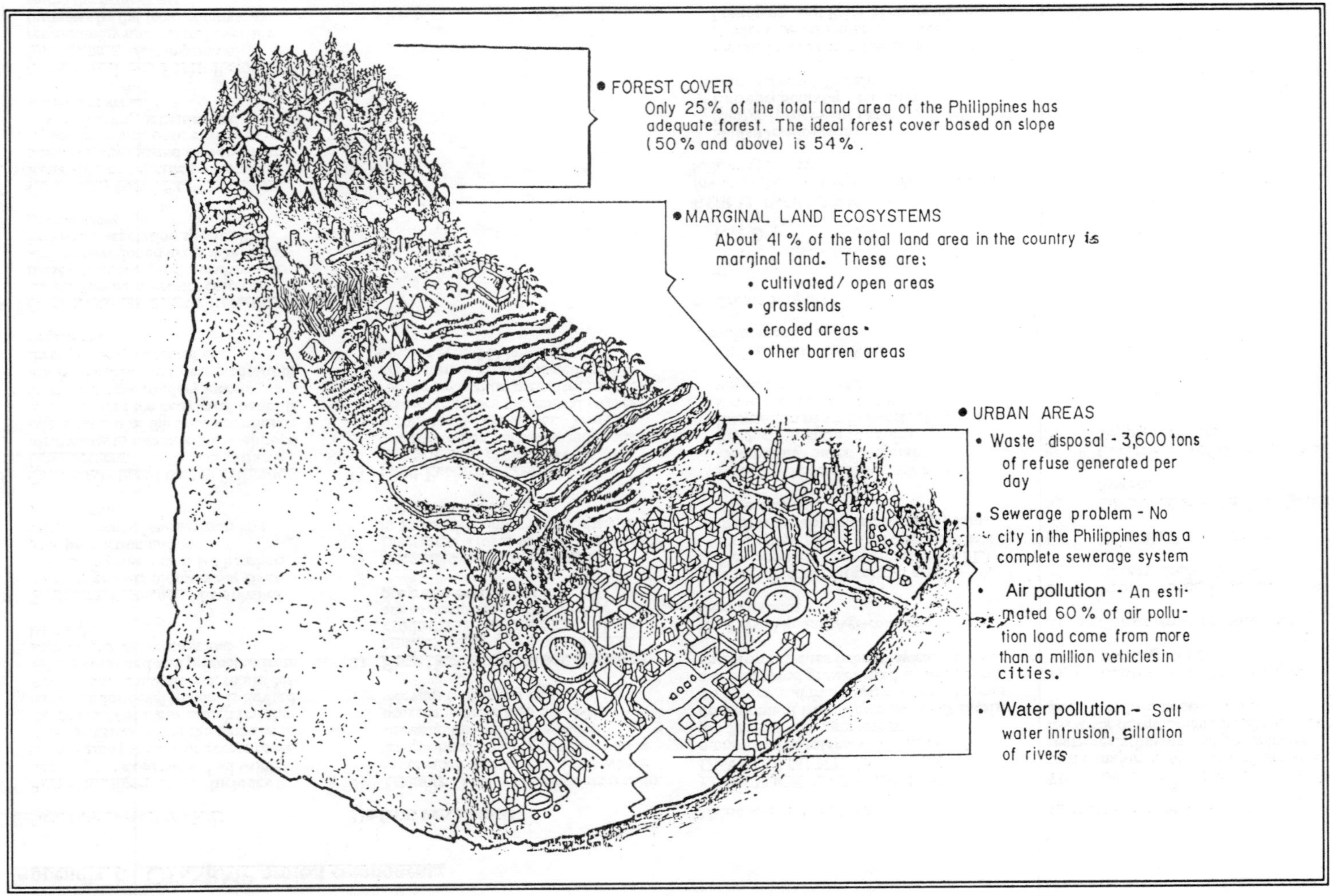

- FOREST COVER
  Only 25% of the total land area of the Philippines has adequate forest. The ideal forest cover based on slope (50% and above) is 54%

- MARGINAL LAND ECOSYSTEMS
  About 41% of the total land area in the country is marginal land. These are:
  - cultivated/open areas
  - grasslands
  - eroded areas
  - other barren areas

- URBAN AREAS
  - Waste disposal - 3,600 tons of refuse generated per day
  - Sewerage problem - No city in the Philippines has a complete sewerage system
  - Air pollution - An estimated 60% of air pollution load come from more than a million vehicles in cities.
  - Water pollution - Salt water intrusion, siltation of rivers

Fig. 2. CBENRMP in the context of national and regional plans.

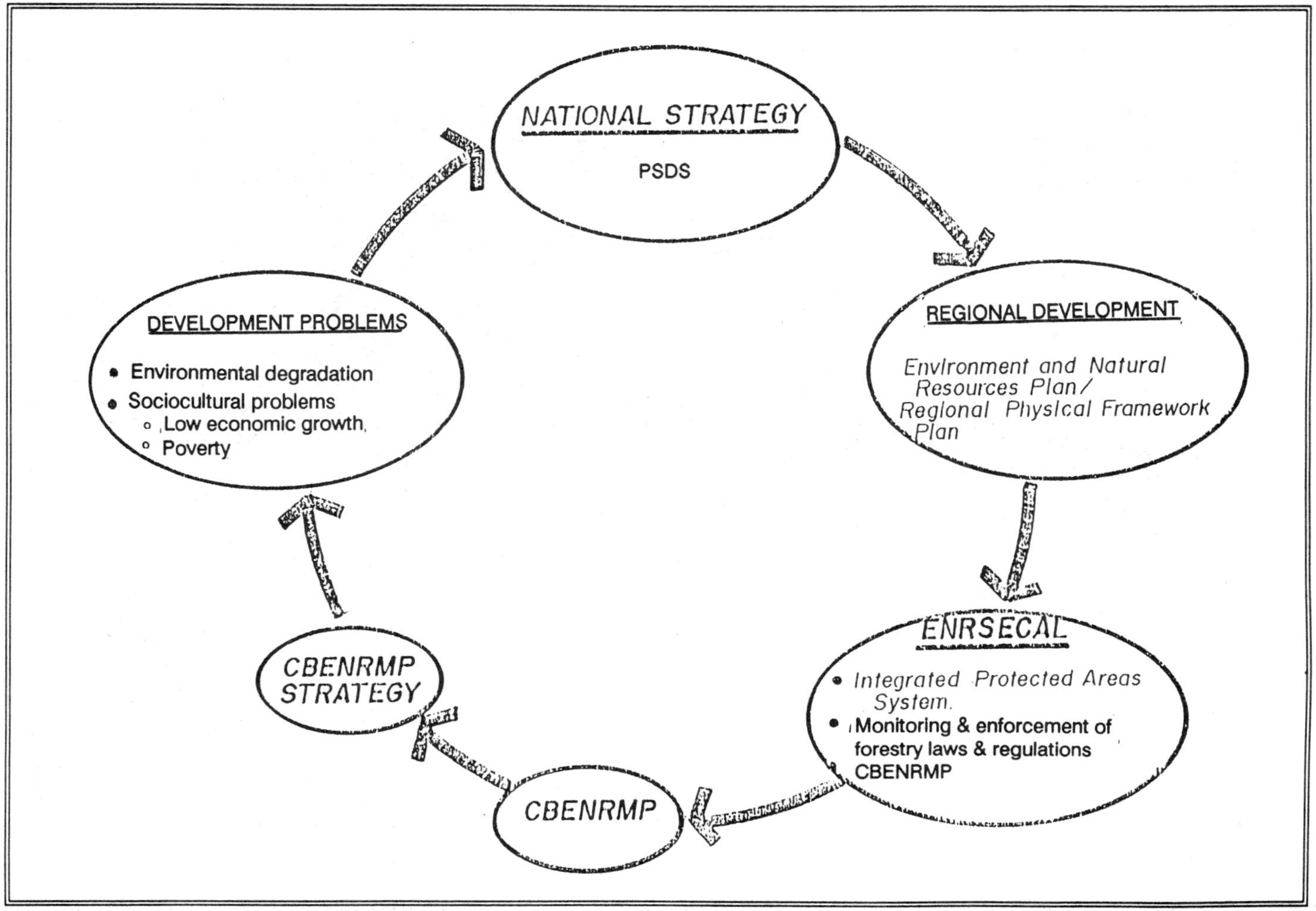

Fig. 3. **CBENRMP sites in the Philippines.**

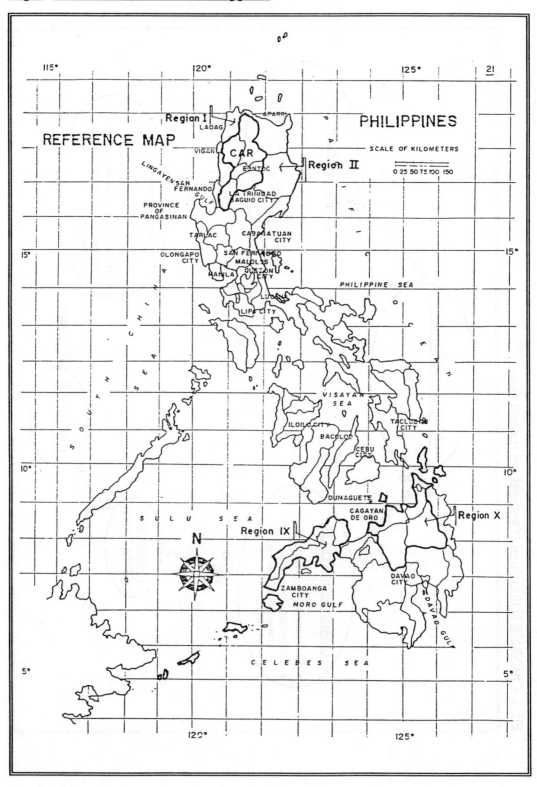

Fig. 4. Map of the Philippines.

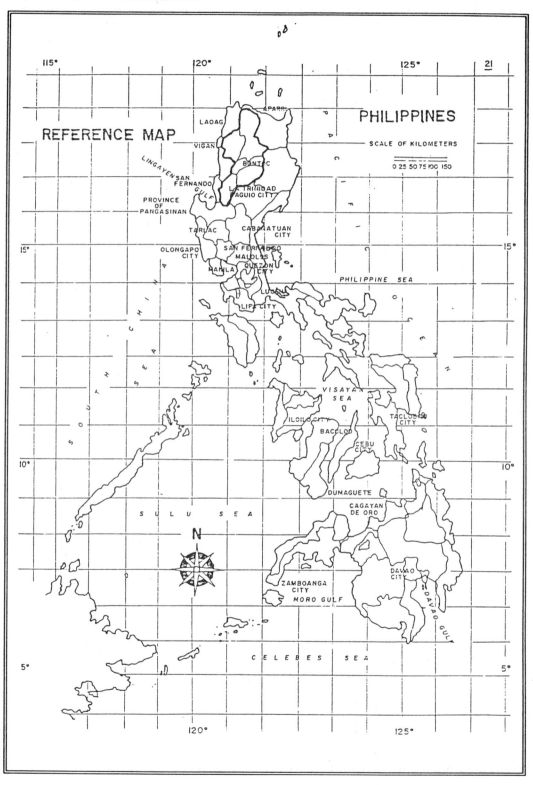

Fig. 5. **Map of Luzon**.

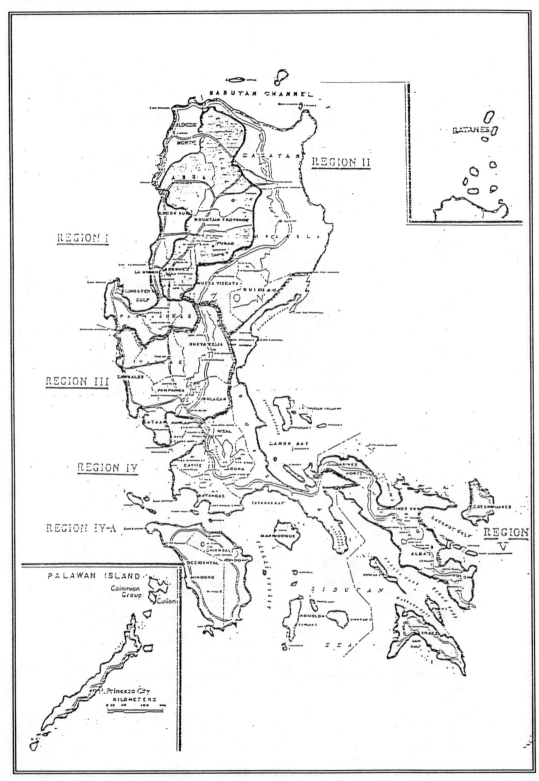

Fig. 6. **Map of Cordillera Administrative Region**.

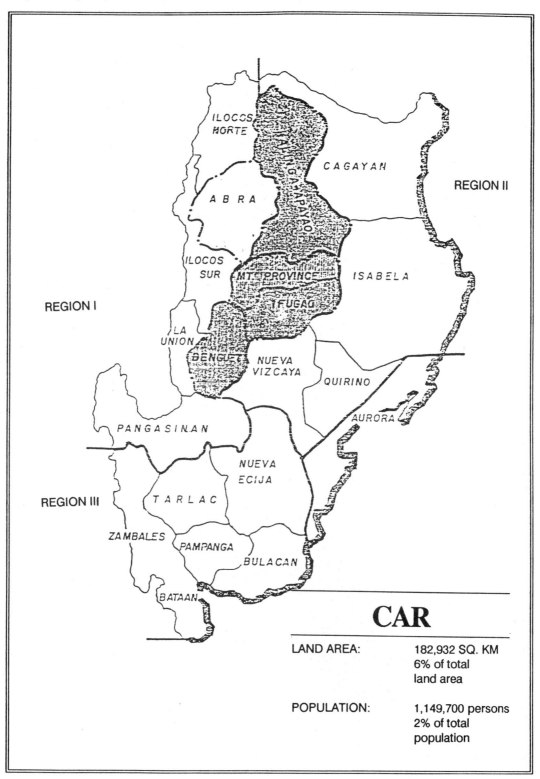

ILOCOS NORTE

CAGAYAN

REGION II

ABRA

KALINGA APAYAO

ILOCOS SUR

MT. PROVINCE

ISABELA

REGION I

IFUGAO

LA UNION

BENGUET

NUEVA VIZCAYA

QUIRINO

AURORA

PANGASINAN

NUEVA ECIJA

REGION III

TARLAC

ZAMBALES

PAMPANGA

BULACAN

BATAAN

CAR

LAND AREA:	182,932 SQ. KM 6% of total land area
POPULATION:	1,149,700 persons 2% of total population

Fig. 7. CAR slope map.

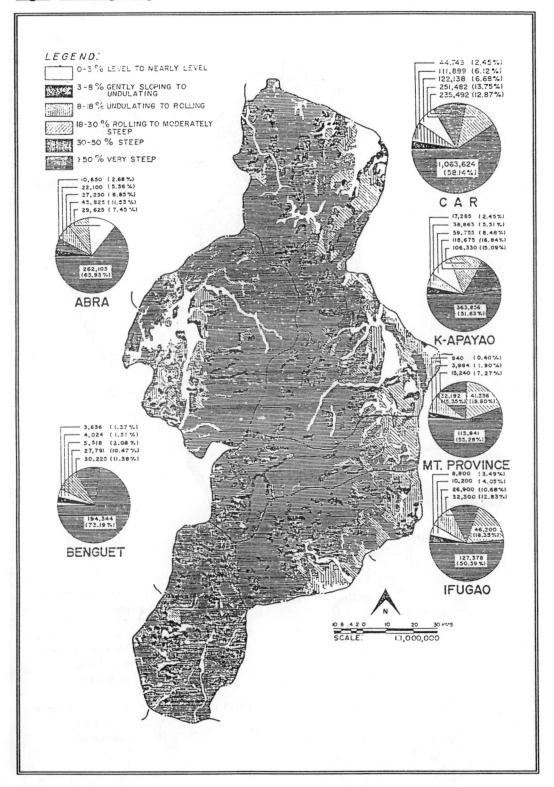

Fig. 8. CAR forest lands and alienable and disposable areas.

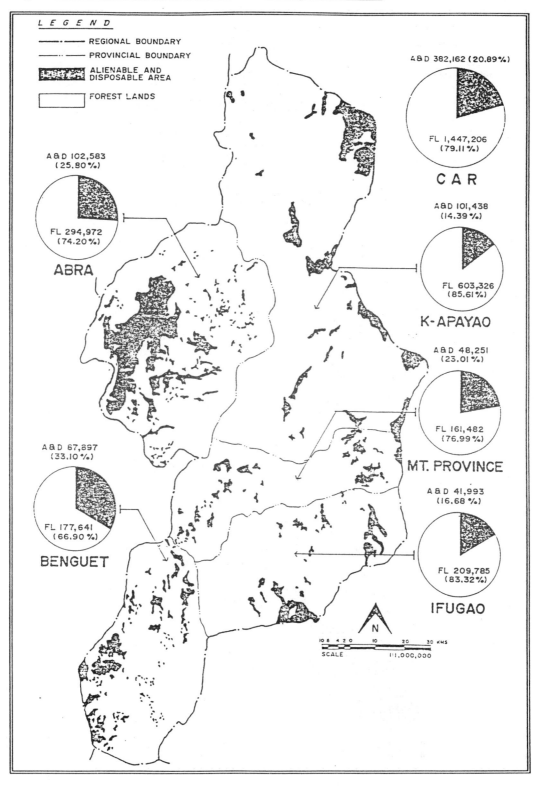

LEGEND

REGIONAL BOUNDARY
PROVINCIAL BOUNDARY
ALIENABLE AND DISPOSABLE AREA
FOREST LANDS

A&D 382,162 (20.89%)
FL 1,447,206 (79.11%)
CAR

A&D 102,583 (25.80%)
FL 294,972 (74.20%)
ABRA

A&D 101,438 (14.39%)
FL 603,326 (85.61%)
K-APAYAO

A&D 48,251 (23.01%)
FL 161,482 (76.99%)
MT. PROVINCE

A&D 67,897 (33.10%)
FL 177,641 (66.90%)
BENGUET

A&D 41,993 (16.68%)
FL 209,785 (83.32%)
IFUGAO

N

10 8 4 2 0   10   20   30 KMS
SCALE   1:1,000,000

Fig. 9. CAR vegetative cover map.

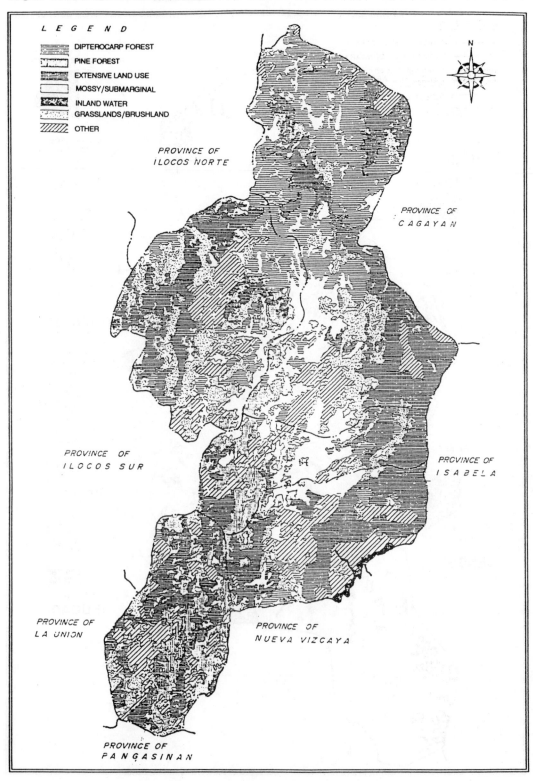

Fig. 10. **CAR land use opportunity map.**

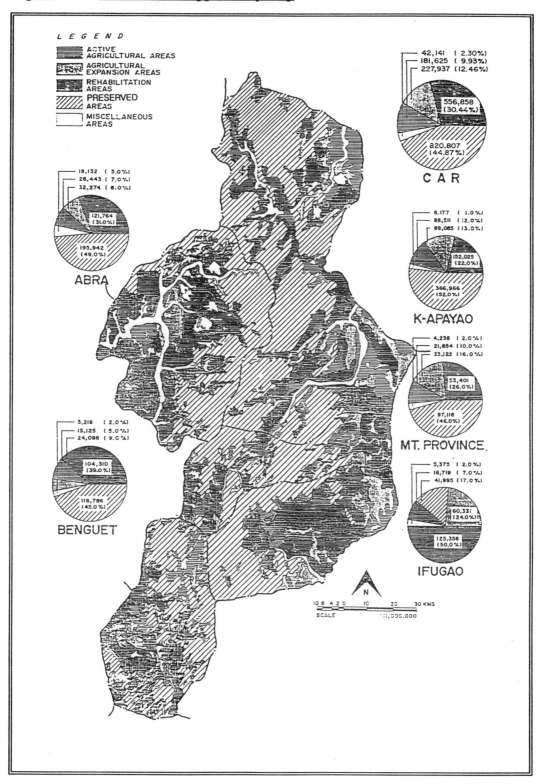

Fig. 11. **CAR erosion map.**

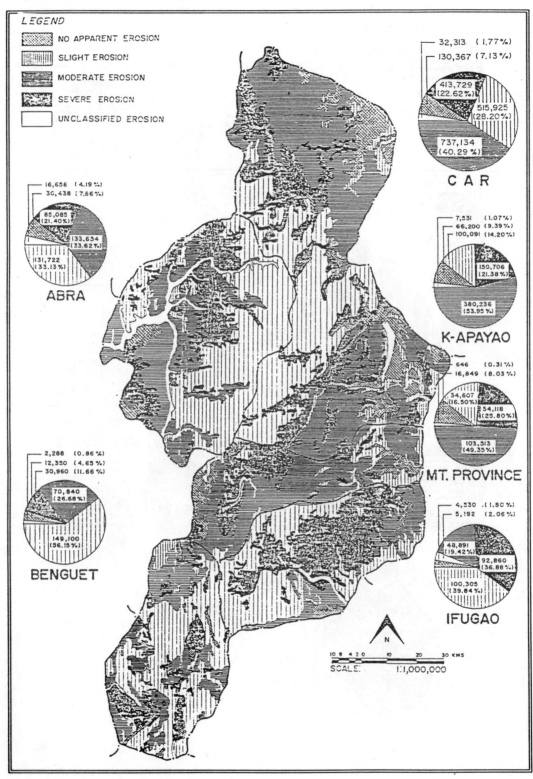

Fig. 12.   **CAR river systems map.**

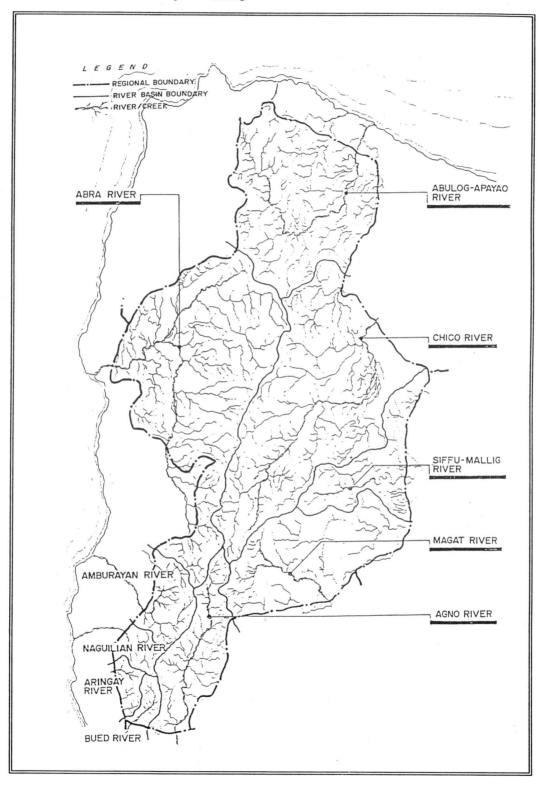

Fig. 13. CBENRMP development direction.

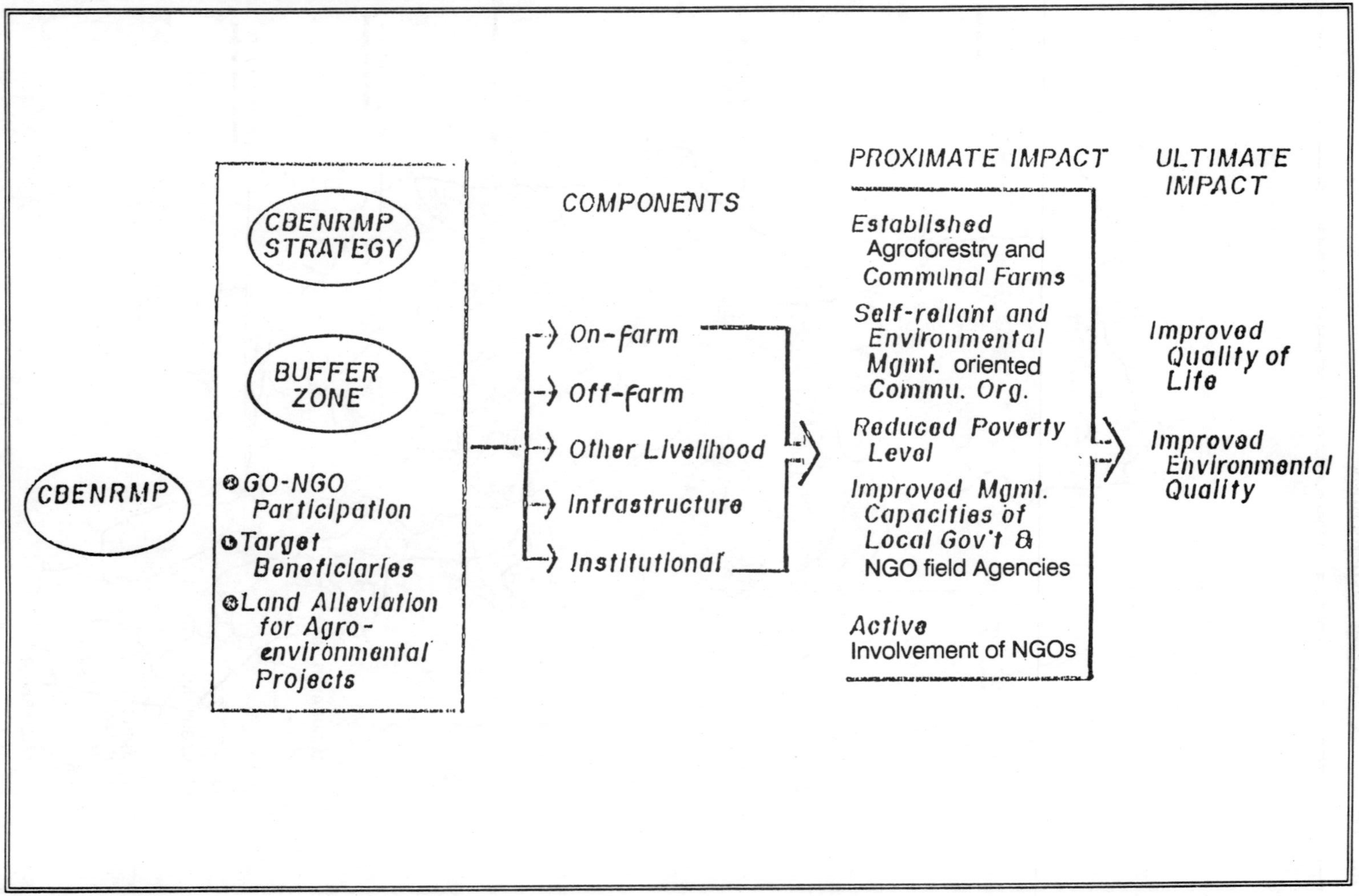

Fig. 14.   Project preparation process.

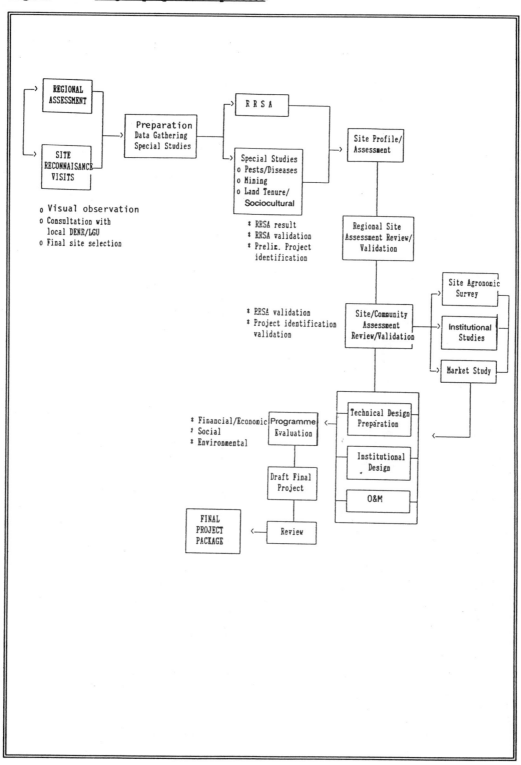

Fig. 15. General process of site selection.

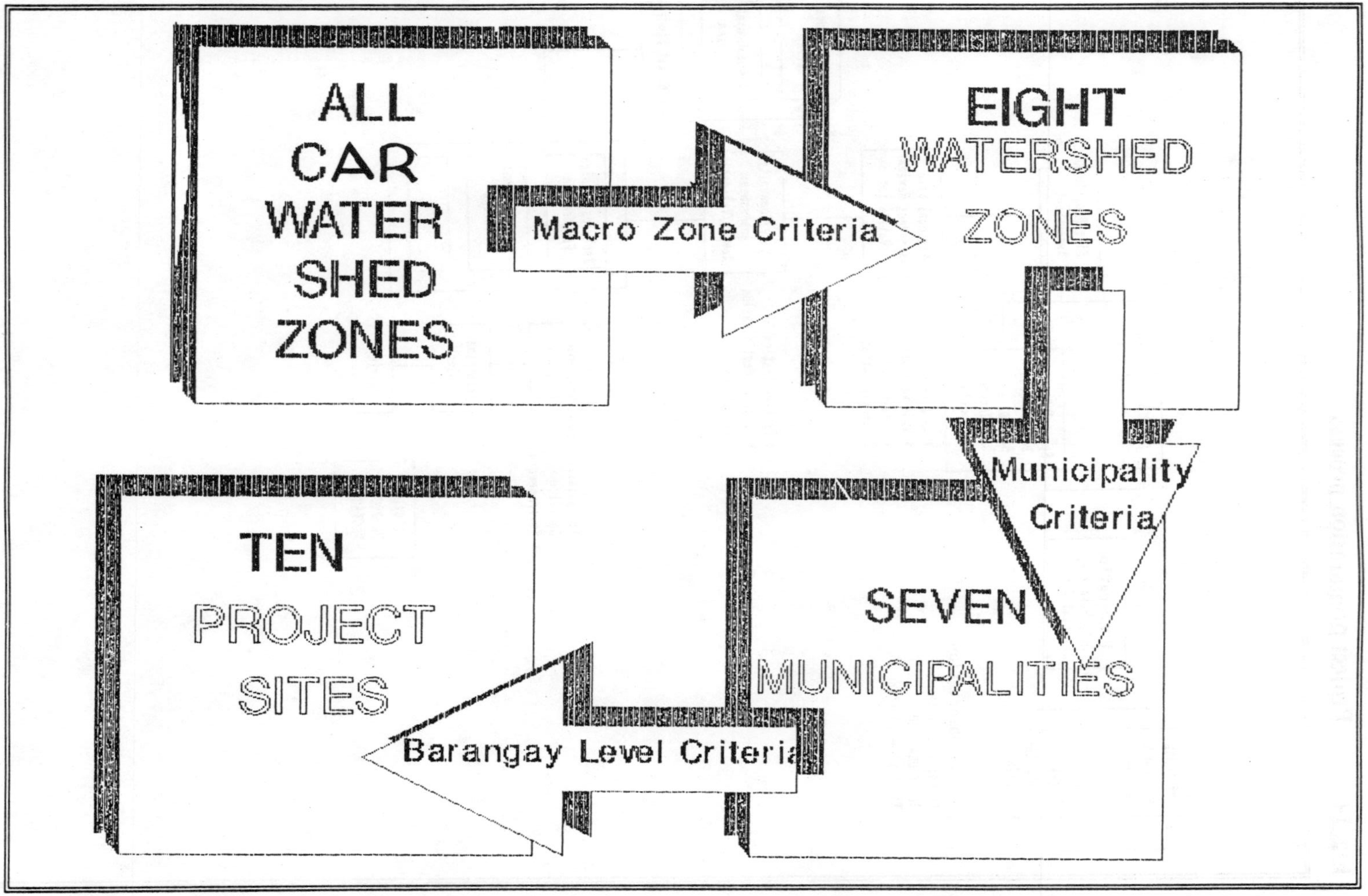

ALL CAR WATER SHED ZONES

Macro Zone Criteria

EIGHT WATERSHED ZONES

Municipality Criteria

SEVEN MUNICIPALITIES

Barangay Level Criteria

TEN PROJECT SITES

Fig. 16. **CBENRMP management**.

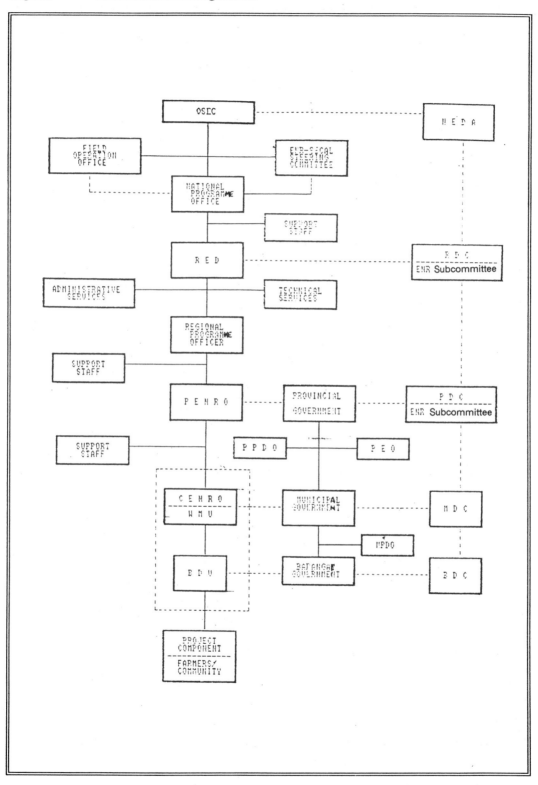

Fig. 17. **General watershed management model for CBENRMP.**

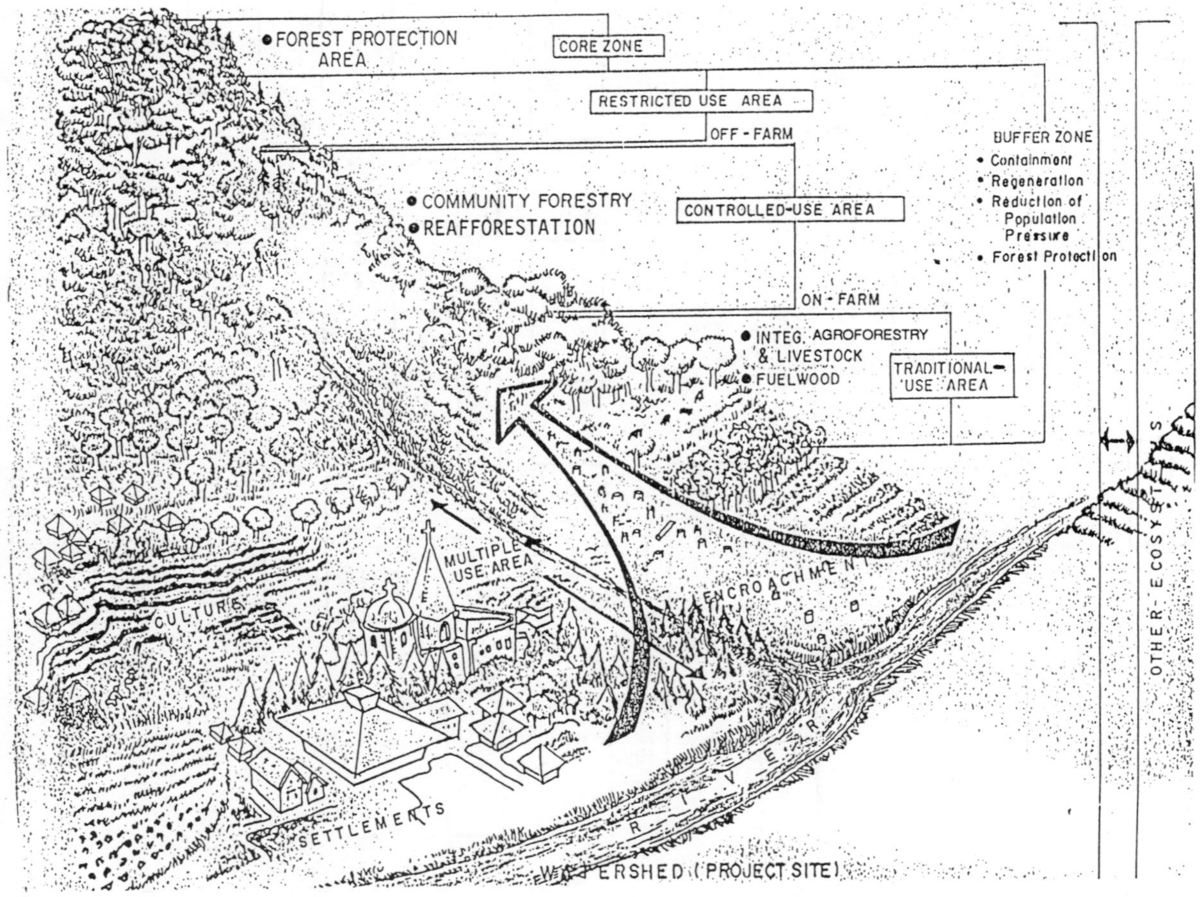

COUNTRY PAPER: PHILIPPINES (2)

Leonardo D. Angeles
Executive Director and
Secretary,
Board of Directors
Philippines Wood
Products Association
Metro Manila,
Philippines

1.   Introduction

The Environment in Crisis

The serious dangers arising from the misuse or abuse of the environment which society now faces were much earlier predicted by people in various scientific disciplines. Theirs were the silent voices of warning in the wilderness, as society then, not yet affected by the slow-building up of these dangers, was not concerned. Now the environmental crisis is real and can become worse. Present-day global society is concerned.

Today society's concern about the environment is not only focused on the appreciation of the cause and effect, for example, of ozone depletion, CO_2 build-up and global warming, acid rain, air and water pollution, erosion and siltation, land misuse, deforestation and conversion, overpopulation and uncontrolled migration, etc. but also on the application of initiatives or actions addressing the mitigation if not reversal of environmental degradation. A focus of intense concern is the enormous and rapid degradation or depletion of tropical rainforest that has been attributed to conversion into subsistence farming (63%), permanent agriculture (16%), cattle ranching (6%), and infrasturcture (1%) as well as to destruction by fuel wood gathering (8%) and logging (6%) (Myers, 1983, FAO, 1987). Global concern for man and the environment was first highlighted in the 1972 UN Conference on the Human Environment (Habito, 1991) which was organized to bring industrialized and developing nations together to discuss the rights of the human family to a healthy, productive, and sustainable environment.

Initiatives for Environmental Integrity, Protection, and Preservation

Initiatives and actions for environmental integrity, protection, and preservation have been taken as a responsibility by various international, regional, and state organizations. The UN has several agencies organized to address environmental problems and to take positive actions including UNEP, UNDP, WHO, UNPF, FAO, and ESCAP (for abbreviations see Appendix 1). Supportive of the development and implementation of environmental policies, programmes, and projects are the multi- and regional funding institutions, such as WB-IMF, EC, and regional development banks in or for Africa, Asia, the Caribbean, and Inter-America, which together with

CEC, OAS, UNDP, and UNEP in New York in 1980 issued the Declaration of Environmental Policies and Procedures Relating to Economic Development (Rees, 1986). Similarly, there are bilateral agencies like JICA and OECF of Japan, USAID, CIDA, and SIDA and ITTO, WWF, and GEF, the last organized by 25 countries during a meeting in Paris in 1990 to establish a fund facility to be implemented jointly by the WB, UNEP, and UNDP (Umali, 1991). Also, in many countries, aside from the governmental institutional agencies, NGOs and PVOs address various environmental concerns.

Appreciation of, concern for, and actions to protect the environment have become common worldwide. There is an urgent need to manage the global environment as well as natural resources integratedly and wisely since they are the mutual heritage of all.

Environmental Concern and Management: The Philippine Experience

Environmental degradation is not a present-day phenomenon in the Philippines nor is concern for and awareness of a growing scarcity of environmental resources a recent development. In the past there were warning voices about the degradation of the country's environment and despoilation of its natural resources, particularly the forest. The advent of the Aquino administration in 1986 signalled a more comprehensive approach toward environmental management. The year 1988 was the beginning of intense public awareness of, and participation in, issues pertaining to the environment and its degradation, and natural resources and their depletion.

The raison d' etre of establishing the Department of Environment and Natural Resources was to be able to provide a holistic approach to the management of the environmental and natural resources by the state. That the pursuit of economic development will consider the preservation of the integrity of the environment and the conservation and wise utilization of natural resources on a sustainable basis for intergenerational ends is the principle and paradigm of managing the environment and natural resources.

Initiatives for Environmental Management

Both short- and long-term programmes and actions of the state are directed toward cleaning the air and the water, protecting the land and the seas, and conserving the resources thereon, and involving various sectors of society in these programmes and endeavours. These programmes and actions are guided by policy reforms having three major thrusts (Factoran, 1990):

1) The first addresses the needs and interests of small indigenous resource users as well as the general public.

2) The second focuses on policies that will allow private industries to utilize the country's natural resources in a more sustainable fashion.

3)   The third includes policy measures that will help rehabilitate the environment.

In pursuance of these agendas specific to the environment and natural resource management, the Philippines is fortunate to being assisted, inter alia by the WB, pledges to PAP started in Tokyo, the ADB and OECF of Japan, USAID, and WWF through its debt-to-nature swap programme.

2.   Management of Forest Resources

Forest Area Changes

Figure 1 illustrates the changes in the dipterecarp forest during the past 50 years:

Figure 1.

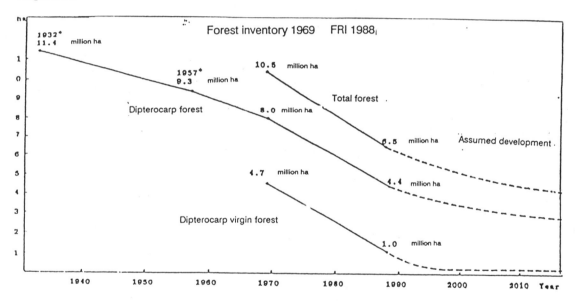

Source: Natural Forest Resources of the Philippines. Philippine-German FRIP, DENR, 1988.
* Figures given by Serevo, Asiddao, and Reyes, 1962.

Between 1969 and 1988, the total forest area decreased by 4 million ha, corresponding to a forest loss of 210,000 ha/yr as in Table 1. The decrease was due to the conversion into residual forest and other land use (Philippine-German NRIP, 1988).

Table 1. Decrease in forest area, 1969-1988.

FOREST TYPE	Area 1969	Area 1968	Annual Rate of Change 1969-88	
	(HA)	(HA)	(HA)	(%)
All forests	10,456,500	6,460,600	-210,300	-2.5
Dipterocarp forests	8,018,900	4,401,100	-190,400	-3.1
Dipterocarp old growth forest	4,656,600	899,300	-192,900	-7.8

Source: Ibid.

Present Extent and Use of Forest vis-á-vis Overall Land Use

Under the 1986 Constitution of the Philippines all lands of the public domain, waters, minerals, coal, petroleum, and other mineral oils, all forces of potential energy, fisheries, forests, or timber, wildlife, and other natural resources are owned by the state. Lands of the public domain are classified into agricultural, forest or timber, mineral lands, and national parks (Constitutional Commission, 1986). The present extent of the forest vis-á-vis the overall land use is shown in Table 2 and illustrated in **Fig. 2.**

As noted in Table 2, forest land constitutes about 53% of the total land area of the country with the balance consisting of intensive agricultural and other uses (**Fig. 3**).

Access to the forest to conduct logging used to be through license, notably timber license agreements (TLAs). While existing TLAs are allowed to continue up to their expiry period, the new modes of access provided for under the 1986 Constitution are through co-production, joint ventures, or production-sharing agreements between the state and Filopino citizens or corporations at least 60% of whose capital is owned by such citizens. The new modes of access are yet to be implemented by an enabling act or law. In the meantime, existing TLAs are still located within parts of the forest as indicated below in Table 3.

Table 2. Extent of forest and land use.

Land use	Area (000 ha)	%
Forest		
Old growth dipterocarp	984	3.3
Second growth dipterocarp	3,456	11.5
Pine	238	0.8
Mossy/marginal	1,413	4.7
Plantation	483	1.6
Extensive land use		
Brushlands	2,459	8.2
Large-scale grasslands	1,543	5.1
Other extensive	6,595	22.0
Intensive land use	11,787	39.3
Waterways, built up, etc.	923	3.1
Total:	30,000	100.0

Source: Master Plan for Forestry Development, DENR 1990.

Table 3. Extent of forest resources under TLAs.

Region	No. of TLAs (1)	Area under TLAs (1) (ha)	Area of forest land (2) (ha)	Total land area (2) (ha)	Annual AC for TLAs (1) ( )
1	1	69,000	1,112,215	2,156,045	No data
2	18	586,000	2,574,080	3,640,300	933,590
3	-	-	771,174	1,623,092	-
4	9	383,973	2,546,520	4,756,016	301,261
5	-	-	541,189	1,763,249	-
6	-	-	613,929	1,002,311	-
7	-	-	935,919	1,495,142	-
8	1	26,600	1,119,454	2,143,149	26,002
9	7	177,363	1,003,189	1,848,514	208,674
10	9	425,942	1,765,426	2,832,774	330,780
11	14	568,975	1,956,839	3,169,275	650,071
12	4	116,215	1,342,738	2,329,323	148,080
Total	63	2,354,068	16,282,672	28,759,190	2,598,458

Sources (1) List of Existing TLA as of Dec. 31, 1990.
  (2) 1990 Philippine Statistical Yearbook, NSCO

The area covered by TLAs is only 8% of the total area of the country, 15% of the entire forest land, and 35% of total forest cover (**Fig. 4**).

Of the 17 million upland people, 9 million live around forest land and 8 million are within the forest land or forest practicing shifting cultivation or slash-and-burn-agriculture (**Fig. 5**). The rest of the forest land is devoted to pasture or grazing, plantation agriculture, and others.

Economic and Social Significance of Forests

Forest resources, especially timber, through the wood-based industry supported the economy recovery or reconstruction of the country from the end of World War II up to the 1970s through export.

Table 4. Contribution of wood industry in exports.
(million US$).

	All exports	Wood products	% share
1950	332.7	10.7	3.2
1960	535.4	91.6	17.1
1970	1,142.2	276.0	24.2
1980	5,787.8	384.0	6.6
1989	7,821.1	173.0	2.2

Over time, the industry's contribution to the economy has decreased as shown together with that of other primary resource-based activities (agriculture, mining, and fisheries). Their total GDP contribution over the last 18 years was on an average of 30%.

Table 5. **Economic contribution of forestry**.

	1970 (million P)	1988
Primary resource-based contribution to gross value added	15,800 (100%)	29,400 (100%)
Forestry only	3,170 (20%)	1,998 (7%)
All manufacturing contribution to gross value added	18,812 (100%)	26,180 (100%)
Wood only	1,181 (10%)	1,309, (5%)

Source: MPFD, 1990.

Aside from economic contributions, the forest-based and -related industries in 1989, for example, provided about 274,000 full-time jobs in wood production and conversion, reafforestation and industrial forest plantation , etc., and another 18,500 persons worked in government, education, and research services.   In addition, forest land provides livelihood through farming activities to an estimated 1.2 million households.   Thus, about 9 million people are dependent on the forestry sector (MPFD, 1990).

Close to 7 million or 65% of the country's households use biomass fuels (mainly fuel wood) for cooking and about 64% of such wood comes from forest land. This was estimated in 1990 to be 31 cum of fuel wood in households and 8 million cum in industries (MPFD, 1990).

Since the take-off of the wood-based industry following World War II much forested land was opened up to settlements of people thus preventing overpopulation in other parts, providing land for the landless, and diffusing social tension due to land conflict and tenancy.   The push toward reconstruction, economic development, and resettlement the primary reasons for forest depletion and deforestation.   Thus the present issues of: environmental problems, the mode to stop logging and by extension the closing of the wood-based industry, and the restructuring of forest policies preoccupy the country.

Causes of Forest Destruction

The loss of the country's forest may be attributed to two interrelated factors: overpopulation (**Figs. 6 and 7**) and poverty, especially in the uplands. In 1934 or over half a century ago, the country had about 17 million ha or 57% of its land in forest with the primary forest covering 11 million ha. Today there are only about 6.5 million ha of forest (excluding brushland) with less than 1 million ha in a primary state (MPFD, 1990). Contributing to massive forest destruction in the later decades has been the pressure from an increasing population in need of land for farming and wood, the over-exploitation of the timber resource, (**Fig. 8**) the inadequacy of forest policy and development, management and conservation efforts, and ineffective enforcement of environmental and forestry laws and regulations.

The rapid depletion of forest brought in its trail erosion and sedimentation of rivers, reservoirs, and seas, flashfloods, and drought, mainly because of improper practices carried out in the uplands. This has laid to virtual wasteland an area close to 11 million ha (MPFD, 1990), which has caused a declining wood supply for the industry.

Following Myers' (1984) definitions, the loss of the Philippine forest (**Fig. 9**) is because of conversion, which includes degradation, impoverishment, destruction, and other changes. Conversion can range from marginal modification to fundamental transformation. Modification may be construed as a result of human intervention whereby the outward appearance, the biological make-up, and ecological dynamics of an original forest undergo discernible change--slight (e.g., selective logging), substantial, or severe (e.g., subsistence farming). Transformation, by contrast, amounts to a basic conversion to make way for permanent agriculture, cash-crop plantations, or pasturelands whereby the forest is totally eliminated and replaced by a man-established ecosystem.

Conversion by modification is happening in the remaining forest land of the country.

Reports and statistics (Tables 6 and 7) will reveal that the rate of conversion (deforestation) of 195,000 ha/yr was mainly due to the transformation of forest land into mostly permanent agricultural land. Accordingly, logging in the mainly forest only causes temporary damage to about 9,000 ha/yr. However, the slash-and-burn agriculture of poor upland dwellers, illegal logging for timber, and fuel wood gathering are the principal cuases of deforestation bordering on the permanent loss of forest.

Table 6. Forest cover and population. 1934 to 1990.

Year	Forest cover (mil. ha)	Forest converted ('000 ha)	Popula- tion ('000)	Forest per capita (ha)	Conversion per capita (ha)	Forest logged ('000 ha)	Logging damage ('000 ha)
1934	17.0		14,412	1.180		27	
1935	16.9	100	14,716	1.148	0.007	31	3
1936	16.8	104	15,027	1.118	0.007	37	3
1937	16.7	108	15,344	1.088	0.007	43	4
1938	16.6	112	15,668	1.058	0.007	40	4
1939	15.5	116	15,999	1.029	0.007	23	4
1940	16.3	120	16,330	1.001	0.007	26	2
1941	16.2	125	16,667	0.973	0.007	30	3
1942	16.1	130	17,012	0.946	0.008	33	3
1943	16.0	135	17,364	0.919	0.008	36	3
1944	15.8	140	17,722	0.892	0.008	40	4
1945	15.7	145	18,089	0.866	0.008	43	4
1946	15.5	150	18,463	0.840	0.008	47	4
1947	15.4	155	18,844	0.815	0.008	50	5
1948	15.2	160	19,234	0.790	0.008	53	5
1949	15.0	165	19,791	0.760	0.008	57	5
1950	14.9	170	20,364	0.730	0.008	60	6
1951	14.7	175	20,953	0.701	0.008	63	6
1952	14.5	180	21,560	0.673	0.008	63	6
1953	14.3	185	22,184	0.646	0.008	63	6
1954	14.1	190	22,826	0.619	0.008	63	6
1955	13.9	195	23,486	0.594	0.008	63	6
1956	13.7	200	24,166	0.569	0.008	72	7
1957	13.5	206	24,866	0.544	0.008	77	8
1958	13.3	213	25,585	0.521	0.008	80	8
1959	13.1	221	26,326	0.498	0.008	91	9
1960	12.9	230	27,088	0.475	0.008	105	11
1961	12.6	240	27,922	0.452	0.009	110	11
1962	12.4	250	28,782	0.430	0.009	113	11
1963	12.1	260	29,668	0.409	0.009	128	13
1964	11.9	270	30,581	0.387	0.009	109	11
1965	11.6	280	31,523	0.367	0.009	103	10
1966	11.3	288	32,494	0.347	0.009	134	13
1967	11.0	294	33,494	0.328	0.009	130	13
1968	10.7	298	34,525	0.310	0.009	185	18
1969	10.4	300	35,588	0.292	0.008	191	19
1970	10.1	300	36,684	0.275	0.008	182	18
1971	9.8	299	37,703	0.260	0.008	177	18
1972	9.5	297	38,751	0.245	0.008	140	14
1973	9.2	294	39,827	0.231	0.008	174	17
1974	8.9	290	40,934	0.218	0.007	170	17
1975	8.6	284	42,071	0.205	0.007	187	19
1976	8.4	276	43,252	0.193	0.006	145	15
1977	8.1	264	44,466	0.182	0.006	133	13
1978	7.8	248	45,714	0.171	0.005	121	12
1979	7.6	230	46,997	0.162	0.005	113	11
1980	7.4	210	48,316	0.153	0.004	135	13
1981	7.2	190	49,540	0.145	0.004	139	14
1982	7.0	170	50,797	0.139	0.004	117	12
1983	6.9	152	52,088	0.132	0.003	114	11
1984	6.8	136	53,410	0.126	0.003	107	11
1985	6.6	122	54,762	0.121	0.003	98	10
1986	6.5	110	56,142	0.116	0.002	90	9
1987	6.4	100	57,547	0.112	0.002	106	11
1988	6.3	94	58,976	0.107	0.002	97	10
1989	6.2	90	60,426	0.103	0.001	50	5
1990	6.1	88	61,894	0.099	0.001	35	4

Notes:

1. Forest cover interpolated from five data points: 1934 - old maps cited by Revilla, 1969 - FAO nationwide inventory, 1976 and 1980 - interpretation of LANDSAT photos, 1987 - nationwide inventory.

2. Forest converted is first derivative of forest cover curve.

3. Population is interpolated from census data.

4. Forest logged is based on log production with 50% allowed for unreported log production.

5. Logging damage is assumed at a high 10% of forest logged.

Source: MPFD, Main Report. TA 933 PHI ADB/FINNIDA. DENR. June 1990

Table 7. Classified and unclassified land area(in million ha).

| Year | Classified A & D | | Classified Forest land | | Unclassified | |
	Area	%	Area	%	Area	%
1955	10.17	35.00	4.00	14.00	15.43	51.00
1965	12.36	41.00	7.73	26.00	9.91	33.00
1975	12.97	43.00	9.11	30.00	7.90	27.00
1985	14.66	49.00	14.02	47.00	1.32	4.00
1990	14.12		15.00		0.88	

Source: Philippine Forestry Statistics (various editions). BF, BFD. PMB.

Response to Forest Destruction

The response of the government to halt forest destruction has been directed mainly against the visible ethical loggers (illegal activities are considered difficult to control or regulate) who are perceived by society in general as the principal culprits of forest destruction. This is contrary to the findings of Myers (1984) that by far the number one factor in the disruption and destruction of tropical forests is the speculative small-scale farmer or the obligate slash-and-burn cultivator who has no land, no gainful work or no prospect of it, no cash income, and a declining real income and cultivating half to one hectare of forest land. Both speculative and obligate farmers, however, maintain a low profile and the government ignores them as not existing at all. The illegal logger, in turn, uses stealth and corruption. With little capital and no taxes, he takes the gamble of being a fugitive from the law but making huge profits from his illegal activity if not apprehended. Thus, most forest policies and regulations of the government are against legitimate loggers, including:

1) cancellation or suspension of TLAs;

2) implementation of the ban on export of logs and lumber;

3) banning logging in primary forest;

4) reduction of annual cutting from the forest;

5) rigorous licensing and regulation of logging and wood processing;

6) increasing forest taxes many fold; and

7) threat of a total commercial logging ban.

On the positive side, however, the government, for the first time, has been successful in key areas of environmental and natural resource rehabilitation and protection by:

1) implementing a comprehensive envrionmental and forest development plan called the Master Plan for Forestry Development;

2) Increasing annual reafforestation goals together with the private sector such as TLA holders through the National Reforestation Programme; and

3) introducing innovative funding schemes for environmental and resource development purposes.

Insofar as the NGOs are concerned, the most significant thrusts toward environmental protection have been:

1) The organization of ecological "bandwagons," e.g., Green Coalition and Earth Savers, headed by Senate protagonists Sen. Mercadeo (total logging ban advocate) and Sen. Alvarez (selective logging ban proponent) with objectives to raise people's consciousness of environmental degradation and to enact proenvironment laws;

2) the development of the uplands involving some 193 organizations from the government, academia and NGOs; and

3) The undertaking of reafforestation under government contract and/or tree planting in urban centres.

The response of the wood-based industry, through its association (PWPA) has been as follows:

1) the establishment of industrial tree plantations in addition to the protection and development of TLA natural forests;

2) the reafforestation of open and denuded areas in conformity with government regulations;

3) the pursuance of efficiency in logging and wood processing;

4) the initiation of public information and livelihood development, especially in rural areas; and

5) the policing of its rank together with the government's.

3.   Trends in Forest Management

Basis for a Comprehensive Environmental and Forest Resource Development and Management

As a first step, the conceptual framework of the Philippines strategy for sustainable development and management (PSSD) of the environment and by extension also of the resources of the environment and by extension also of the resources of the environment has already been passed by Cabinet Resolution No. 37 on November 29, 1989, which accordingly is seen as the most comprehensive plan document on sustainable development in the Asia-Pacific region (Ganapin, 1991). It provides the goals, objectives, and general strategies for sustainable development. The goal of PSSD is to achieve economic growth with adequate protection of the country's biological resources and their diversity as well as vital ecosystem functions and overall environmental quality.

The basic guiding principles enunciated in the PSSD are:

1) a system-oriented and integrated approach in the analysis and solution of development problems;

2) a concern in meeting the needs of future generations, otherwise termed intergenerational equity;

3) a concern for equity of people's access to natural resources;

4) a concern not to exceed the carrying capacity of the ecosystem but living on the interest rather than on the capital or stock of natural resources;

5) maintenance or strengthening of vital ecosystem functions in every development activity;

6) a concern for resource efficiency;

7) promotion of research on substitutes, recycling, and exploration from revenues derived from the utilization of nonrenewable resources;

8) a recognition that poverty is both a cause and consequence of environmental degradation; and

9) promotion of citizen participation.

Based on the above nine principles, 10 major general strategies were outlined:

1) the integration of environmental considerations in policy and decision making;

2) the proper pricing of natural resources;

3) property rights recognition and reform, especially directed at the 18 million uplanders;

4) the establishment of an integrated protected area system (IPAS), i.e., through the assessment of various biogeographical zones and ecosystems and development of detailed management plans for them;

5) the rehabilitation of degraded ecosystems;

6) the strengthening of residuals management in industry or pollution control (where the implementation of environmental impact assessment would be necessary prior to the issuance of environmental compliance certificates to effect low-waste or no-waste technologies);

7)   integration of population concerns and social welfare in development planning;

8)   the inducing of growth in rural areas;

9)   the promotion of environmental education; and

10) the strengthening of citizen participation and constituency building.

As far as the forest environment and resources are concerned, the recently developed MPFD will serve as the basis of operationalizing the aforementioned strategies. The MPFD has been accepted as the forestry master plan even by donor or funding institutions like the ADB and the USAID's NRMP. A macrolevel plan, it intends to charter the direction the forestry sector should take.   With its indicative scenarios and recommendation for policy reforms and programme components, it is intended to fine tuned at the local level. Its programme consist of the following:

1)   Programme on man and the environment

--   people-oriented forestry;

--   soil conservation and watershed management;

--   integrated protected area system and biodiversity conservation;

--   urban forestry; and

--   forest protection.

2)   Forest management and product development programmes

--   management and utilization of the primary forest, the dipterocarp forest;

--   mangrove and pine forests;

--   forest plantations and tree farms;

--   wood-based industries; and

--   non-wood forest-based industries.

3)   Institutional development programmes

--   policy and legislation;

--   organization, human resources, infra-structure, and facilities;

--   research and development;

--   education, training, and extension; and

--   monitoring and evaluation of activities.

The MPFD is rather an ambitious plan requiring enormous investment to be sourced mainly from aid, grants, donations, and possibly soft loans. To give it a sense of permanency, it is essential that Congress adopts the MPFD or the important programmes therein requiring the passage of laws. The promotion of significant aspects of the conceptual and operational frameworks of sustainable development of the environment and natural resources, especially those which pertain to the continued utilization of the forest resources by the wood-based industry, however, seems unacceptable to the present Congress. The total banning of commercial logging in the forest seems to be the direction of Congress. The pros and cons of a total or selective logging ban in relation to the vision of the MPFD as well as the impact of the same on industry and the national economy are outlined in Tables 8 and 9.

Table 8. Total Versus Selective Logging Ban.

Issue	Current trends scenarios		Master plan scenario
	Total logging ban	selective ban	
Land use			
-Virgin forests	More pouched than reserved*	Reserved	Reserved
-Second-growth forests	More pouched than reserved*	Main production areas	Main production areas
-Production forests	Only plantations	Second-growth & plantations.	Second-growth & plantations
Forest protection and related concerns			
-Protected forests	All forests, limited by gov't funds	Some forests, TLA areas protected	Some forests DENR/TLA/NGO funds
-Forest land managers	DENR alone.	DENR/TLA.	DENR/TLA/ communities.
-Deforestation	High	Low in TLA areas*	Low

-Reafforestation station	Limited by gov't funds	Limited by gov't funds, contribution by private sector high*	Growing private funding
	Mainly gov't effort	Both gov't & private sector	Growing private initiatives.
Wood supply situation			
-From dipterocarps	Only illegal supply	From second-growth forests	From second-growth forests
-From existing plantations	Allowed	Allowed	Allowed
-From new plantations	Medium	Moderately high*	High
-From export	If available, high $ outflow*	If available, low $ outflow*	If available, low $ outflow
-Log transport	Unregulated*	Highly regulated	Regulation to be phased out

Economic and environmental impacts

-Forest industry	On the decline Import dependent if supply is available.*	Continued investments* tied to wood supply	Medium investments, independent
-Wood trade	Wood imports. No exports.	Low imports; low exports.	High exports
-Govt revenue	Very low	Increasing*	High
-Employment	Very low	Prospect increasing*	High

Source: DENR MPFD main report, funded thru TA933 PHI of ADB and FINIDA June 1990.
Note: Asterisks (*) indicate comments by PWPA based on realities, experience, and prospects.

Table 9.

What will a total logging ban do?

Cause Wood Supply Uncertainty

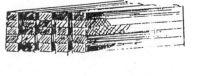

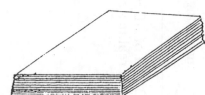

COUNTRY WILL HAVE TO IMPORT 2.5M CUM WORTH US$375 M JUST FOR DOMESTIC CONSUMPTION. THIS CAN BE MET W/O NEED TO IMPORT EVEN BY JUST 30 TLA HOLDERS PRODUCING CURRENT ANNUAL ALLOWANCE CUT OF 3.5 M CUM.

COUNTRY WILL FOREGO ANOTHER US$100 M IN EXPORT OF HIGH VALUE-ADDED FINISHED WOOD PRODUCTS, TOTAL FOREX SPENT/FOREGONE WILL BE NEARLY US$0.5 BILLION WHICH COULD BE BETTER USED FOR OTHER VITAL IMPORTS LIKE OIL.

SINCE TIMBER-EXPORTING COUNTRIES ALSO NEED WOOD PRODUCTS FOR OWN USE, AND RP HAS TO COMPETE WITH OTHER IMPORTERS SUCH AS JAPAN, KOREA, AND TAIWAN WHO HAVE STRONG CURRENCIES. WOOD SUPPLY IS UNCERTAIN. THIS WILL HAMPER RP DEVELOPMENT EFFORTS ESPECIALLY FOR RECONSTRUCTION IN EARTHQUAKE-DAMAGED AREAS.

PRICES OF WOOD PRODUCTS WILL RISE ABOVE REACH OF AVERAGE FILIPINOS. LOS SMUGGLING TO PROLIFERATE GOVT MASS HOUSING PROGRAMME TO SUFFER.

Set back Forest Protection

TOTAL LOGGING BAN DIFFICULT IF NOT IMPOSSIBLE TO ENFORCE AS DENR SECRETARY HIMSELF ADMITS. ONLY 1 DENR GUARD PER 4,000 HA OF FOREST LANDS VS. WOOD INDUSTRY RATE OF 1 EMPLOYEE/GUARD PER 100 HA BAN WILL ADD ANOTHER 2 M HA NOW BEING PROTECTED BY TLA-HOLDERS TO FOREST LANDS UNDER GOVERNMENT RESPONSIBILITY.

WILL ACCELERATE RATHER THAN STOP FOREST DESTRUCTION SINCE FUEL WOOD GATHERERS, KAINGINEROS AND LOG SMUGGLERS WHO HARVEST ALMOST 30M CUM YEARLY WILL OVERRUN LIGHTLY GUARDED FOREST.

WILL FOREGO REAFFORESTATION BY TLA HOLDERS UNDER REFORESTATION TRUST FUND (RTF) PROGRAMME AS WELL AS EMPLOYMENT OPPORTUNITIES FOR PEOPLE IN RURAL AREAS WHERE JOBS BADLY NEEDED.

Cause Economic Dislocations

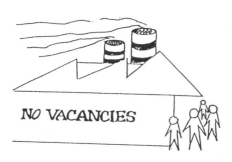

WILL DEPRIVE 200,000 FILIPINOS OF DIRECT MEANS OF LIVELIHOOD FROM PRIMARY, SECONDARY, AND TERTIARY WOOD PROCESSING AND INDIRECTLY AFFECT THEIR ONE MILLION DEPENDENTS. THUS WORSENING COUNTRY'S UNEMPLOYMENT SITUATION.

SINCE MOST ARE IN HINTERLANDS, DISPLACEMENT COULD CAUSE THEM TO JOIN NPA RANKS THUS FOMENTING INSURGENCY FURTHER.

Unjust to responsible TLA holders

FROM MORE THAN 300 IN 1986, NO. OF TLA HOLDERS NOW DOWN TO 65 STILL ACTIVE. TOTAL LOGGING BAN WILL UNJUSTLY PUNISH THESE LAW ABIDING COMPANIES AND DEPRIVE THEM OF FAIR RETURN ON THEIR INVESTMENTS IN INDUSTRY TOTALLING P10 BILLION IN CAPITAL EQUIPMENT ALONE.

COUNTRY WILL ALSO LOSE INDUSTRY CONTRIBUTION TO GDP OF ALMOST P17 BILLION.

Specific Trends in Forest Management

The pursuance of PSSD and the MPFD has created results, as desired by the government, mostly through forestry policy and regulatory measures that have affected the forest-based industries. Among these results are:

1) reductions in the number of TLAs, the area of forest under logging and the production of timber from such forest (Table 10);

2) banning of logging in primary or virgin forest, declaring them the same as IPAS and shifting such logging in residual forest;

3) an increase in forest taxes, by as much as 1,600%, through the imposition of environmental fees and now through the passage of RA 7161;

4) mandating TLA holders to conduct obligatory reafforestation and timber stand improvement programmes starting in 1988 with TLA holders depositing into the Reforestation Trust Fund (TFT) amounts necessary to carry out annual reafforestation/TSI activities (Tables 11 and 12).

5) promulgation of a regulation encouraging the development of industrial forest plantation;

6) banning of the use of high-lead yarding, the operation of more than one logging side, and logging on slopes 50% and over and 1000m and over in elevation; and

7) banning of the cutting of endangered tree species.

For the uplanders in forest land, a livelihood-cum-environmental protection programme, called Integrated Social Forestry (ISF), is being implemented and to secure their tenure over the appropriated areas they are issued certificates of stewardship contract (CSC) for 25 years renewable for another 25 years. In accordance with the MPFD, uplanders may be allowed to engage in community logging of the forest which is designed to replace TLAs.

At no time in the history of forest management has public consciousness, awareness, and involvement been more heightened than today. Complete and total involvement and participation of all sectors of society are essential if a country like the Philippines is to achieve a productive and healthful environment that can be shared by the rest of the community of nations.

Table 10. Periodic Logging Industry Profile: Resources

Years 1955-1990
Area in '000 ha; ha; AAC in '000 cu m

Year/licence	No.	Area	AAC
1955			
TLA			
OTL			
1960			
TLA	27.00		
OTL	1,359.00	3,444.00	4,920.00
1965			
TLA	30.00	1,644.00	2,401.00
OTL	381.00	2,797.00	3,749.00
1970			
TLA	58.00	3,368.00	6,099.00
OTL	267.00	6,267.00	8,436.00
1975			
TLA	171.00	6,921.00	13,736.00
OTL	64.00	2,137.00	3,063.00
1980			
TLA	191.00	6,500.00	13,699.00
OTL	41.00	704.00	902.00
1985			
TLA	148.00	6,093.00	8,903.00
OTL	-	-	-
1990			
TLA	63.00	2,354.00	2,478.00
OTL	-	-	-

Source: Philippine Lumber Industry Statistics (various sources). PLPA

Table 11. **Reafforestation by TLA holders**.

Year	No. TLA holders	Actual Planting (ha)	RTF Amount (million P)
1988	100	24,792	96
1989	103	26,911	248
1990	80	28,393	328
Total		80,096	672

The aspirations of one country, especially one that is underdeveloped or developing, toward these ends deserves the support and cooperation of other countries more blessed materially which may have in the past or even now benefited by the fruits or from the despoilation of that country's environment and natural resources. There is but one global village, this earth, deserving order and care for intergenerational benefits.

Table 12. **Area Reforested annually by the government and private sectors, FY 1960-61 to CY 1989 (ha).**

Year	Grand Total	Government Sector	% Share	Private Sector	% Share
Area plantations in 1989	1,031,113	696,349	68	334,764	32
1989	131,404	89,452	68	41,952	32
1988	64,183	31,226	49	32,957	51
1987	30.811	28,843	72	10,968	28
1986	32,998	24,426	74	8,572	26
1985	24,231	12,684	52	11,547	48
1984	38,935	16,088	41	22,847	59
1983	78,538	42,239	54	36,299	46
1982	63,262	35,201	56	28,061	44
1981	64,541	33,296	52	31,245	48
1980	60,516	39,881	66	20,635	34
1979	79,397	51,858	65	27,539	35
1978	78,425	44,686	57	33,739	43
1977	53,263	33,365	63	19,893	37
1976	31,773	23,228	73	8,505	27
1974-75	15,280	15,280			
1973-74	4,994	4,994			
1972-73	5,787	5,787			
1971-72	4,831	4,831			
1970-71	6,458	6,458			
1969-70	11,801	11,801			
1968-69	7,511	7,511			
1967-68	6,869	6,869			
1966-67	5,327	5,327			
1965-66	7,396	7,396			
1964-65	11,709	11,709			
1963-64	16,822	16,822			
1962-63	11,543	11,543			
1961-62	7,474	7,474			
1960-61	11,543	11,543			

Source: Philippine Forestry Statistics, 1989

REFERENCES

Constitutional Commission. 1986. The Constitution of the Republic of the Philippines.

Factoran, F.S., Jr. 1990. Keynote speech. Consultative Meeting on the Environment. Proceedings of the Consultative Meeting. Makati, Metro Manila.

Ganapin, D.J., Jr. 1991. Update on Philippine strategy for sustainable development. Conference Proceedings. Manila: Philippine Furturistics Society.

Habito, C. F. 1991. Integrating environmental concerns into economic development. Conference Proceedings. Manila: Philippine Futuristic Society.

Master Plan for Forestry Development. Main Report. 1990. TA 993 PHI of ADB and FINNIDA. Quezon City: DENR.

Myers, N. 1984. The Primary Source: Tropical Forests and Our Future. New York, London: WW Norton & Company.

Philippine-German Forest Resources Inventory Project. 1988. Natural Forest Resources of the Philippines. Quezon City: DENR.

Rees, C.P. 1986. Environmental planning and management by Asian Development Bank in its economic development activities. Regional Symposium on Environmental and Natural Resources Planning. Manila: ADB.

Umali, R.M. 1991. Environmental financing: Strategies and approach. Conference Proceedings. Manila: Philippine Futuristic Society.

Appendix 1. List of Abbreviations.

Abbreviation	Explanation/meaning
ADB	Asian Development Bank
CIDA	Canada International Development Assistance
EC or CEC	European Community or Commission of European Community
ESCAP	Educational, Scientific and Cultural Commission for Asia and the Pacific
FAO	Food and Agriculture Organization
FINNIDA	Finland International Development Assistance
GDP	Gross domestic product
GEF	Global Environmental Facility
IMF	International Monetary Fund
ITTO	International Tropical Timber Organization
JICA	Japan International Cooperative Assistance
MPFD	Master Plan for Forestry Development
NGO	Non governmental organization
NRIP	National Resources Inventory Project (RP-German)
OAS	Organization of American States
OECF	Overseas Economic Cooperation Fund
PAP	Philippine Assistance Program
PVO	Private voluntary organization
PWPA	Philippine Wood Products Association
SIDA	Sweden International Development Assistance
UN	United Nations
UNDP	United Nations Development Programme
UNEP	UN Environmental Programme
UNPF	UN Population Fund
USAID	US Agency for International Development
WB	World Bank
WMO	World Meteorological Organization
WHO	World Health Organization
WWF	World Wildlife Fund
cum	cubic meters
ha	hectares
P	Philippine peso
$	US dollars (1 US$ = approx. P27.00)

Figure 2. **Land use and forest types (1,000 ha).**

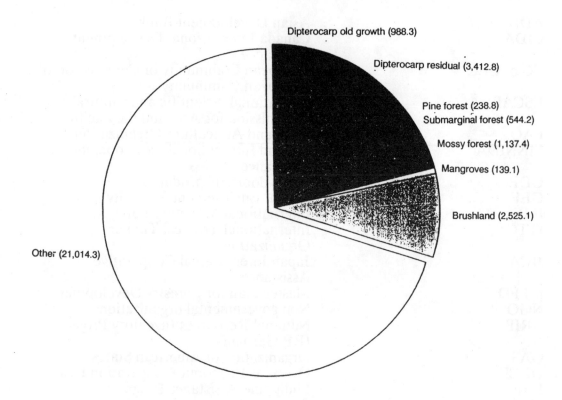

Dipterocarp old growth (988.3)

Dipterocarp residual (3,412.8)

Pine forest (238.8)
Submarginal forest (544.2)

Mossy forest (1,137.4)

Mangroves (139.1)

Brushland (2,525.1)

Other (21,014.3)

Figure 3. Total land area by type (30 million ha).

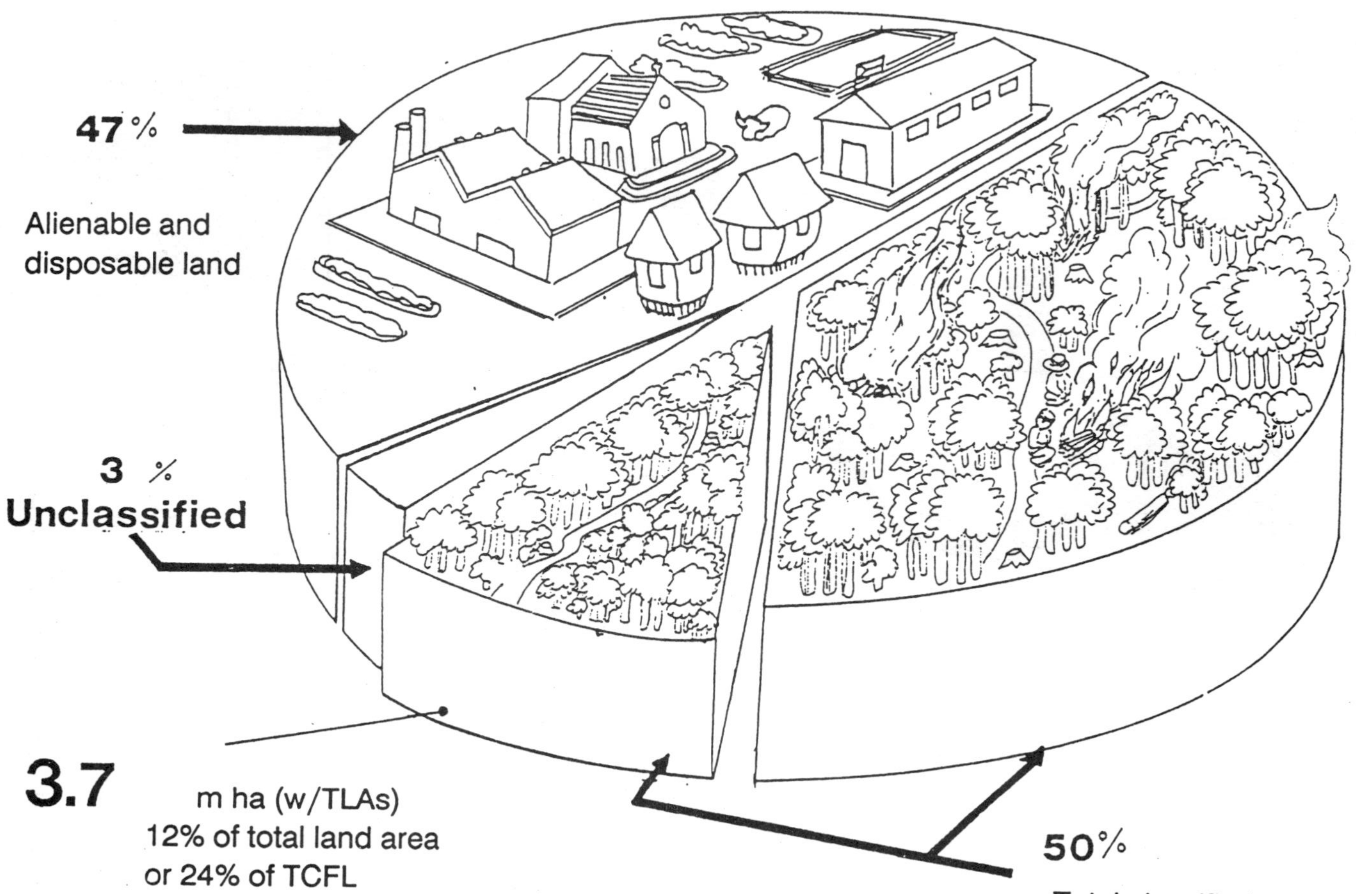

47%
Alienable and
disposable land

50%
Total classified
forest land

3%
Unclassified

3.7 m ha (w/TLAs)
12% of total land area
or 24% of TCFL

SOURCE: Forest Management Bureau, DENR.

— 255 —

Figure 4. Provinces with logging bans vs with TLAs.

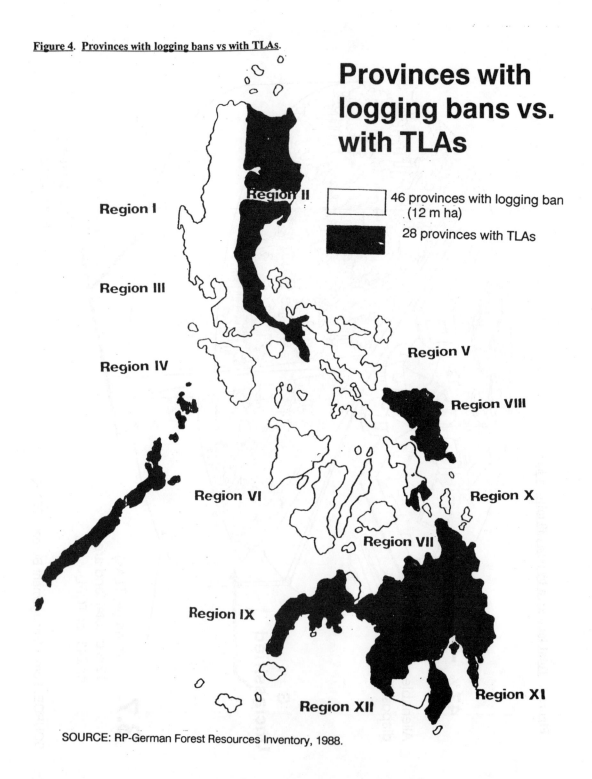

Provinces with logging bans vs. with TLAs

☐ 46 provinces with logging ban (12 m ha)

■ 28 provinces with TLAs

Region I

Region II

Region III

Region IV

Region V

Region VIII

Region VI

Region X

Region VII

Region IX

Region XI

Region XII

SOURCE: RP-German Forest Resources Inventory, 1988.

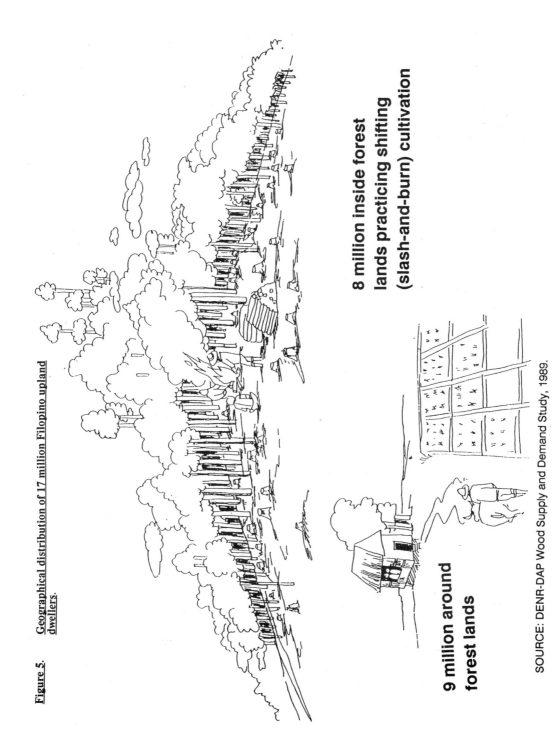

Figure 5. Geographical distribution of 17 million Filopino upland dwellers.

8 million inside forest lands practicing shifting (slash-and-burn) cultivation

9 million around forest lands

SOURCE: DENR-DAP Wood Supply and Demand Study, 1989.

Figure 6. Philippine forest cover and population, 1932 to 1988

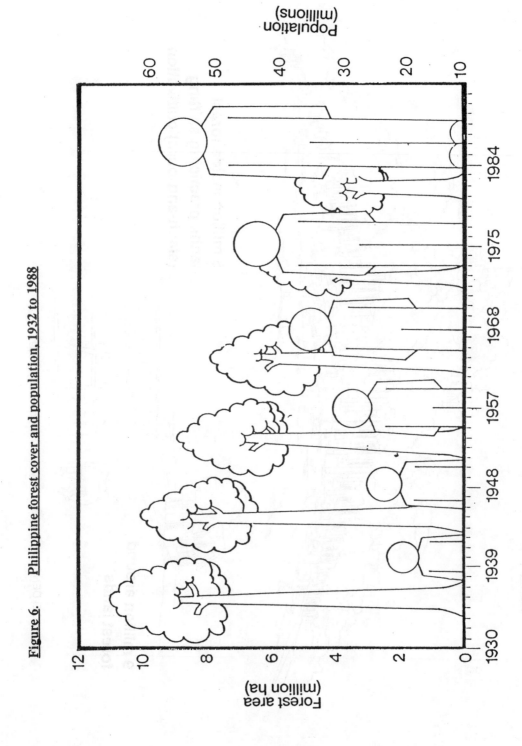

SOURCE: Saarilahti, 1989. Computed from official statistics.

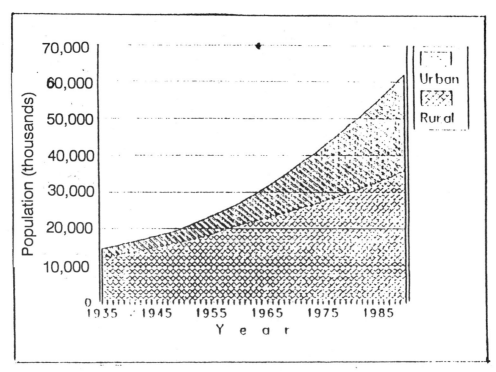

Fig. 7. Population of the Philippines, 1935-1990.

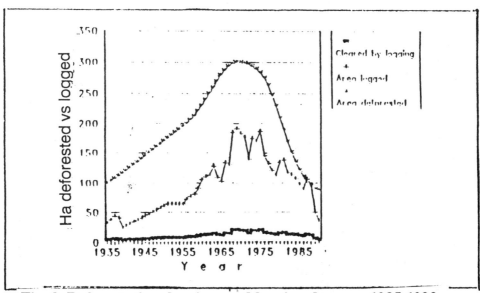

Fig. 8. Deforestation, logging, and logging damage, 1935-1990.

Source: MPFD, DENR, 1990.

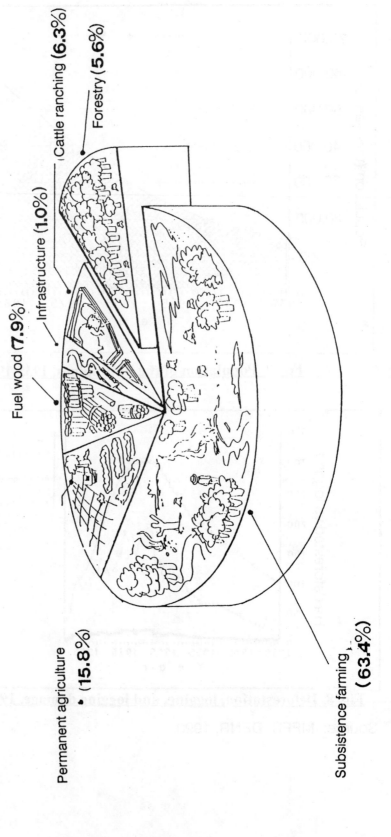

Figure 9. Main causes of rainforest degradation.

(Data from Myers, 1983 and FAO, 1987.)

Cattle ranching (**6.3%**)

Forestry (**5.6%**)

Infrastructure (**1.0%**)

Fuel wood (**7.9%**)

Permanent agriculture (**15.8%**)

Subsistence farming (**63.4%**)

COUNTRY PAPER: SRI LANKA

A.V.N. Silva
Leaf Manager
Ceylon Tobacco Company
Limited
Kandy, Sri Lanka

1.   Introduction

Sri Lanka is an island lying at $5^{0}55'$ to $9^{0}50'$ north of the equator with a tropical climate. The total land extent of Sri Lanka is 6,560,000 ha and the population is around 17 million. The country is divided into three main agroecological zones based on rainfall, ie: 1) the wet zone, with rainfall of 80-120 inches annually received from the northeast monsoon (October to January) and southwest monsoon (May to August); 2) the intermediate zone, with rainfall of 40-60 inches annually from the northeast monsoon and slight precipitation from the southwest monsoon; and 3) the dry zone, with rainfall of 20-40 inches annually only from the northeast monsoon. (Table 1 and **Fig. 1**).

Table 1. Sri Lanka Statistics.

Population	=	17 million		
Total land area	=	6,560,000 ha		
Population distribution	=		Urban	22%
			Rural	78%
Literacy rate	=			87%
Forest cover	=	1900:		70%
		1960:		44%
		1990:		27%
Topography	=		Coastal and northern areas flat	
			Central and southern areas hilly and mountainous	

Where energy is concerned, trees provide 78% of the total requirement, while imported fossil, fuel and hydropower provide 19.3% and 2.7%, respectively. The major user of wood fuel is the domestic homestead (for cooking) which accounts for 82% of the total fuel wood usage, while industries (tea, bricks, roof and floor tiles, bakeries, lime kilns, and tobacco) account for the remaining 18% (**Fig. 2**).

The forest cover of Sri Lanka, which was 44% in the early 1960s, dropped drastically due to large-scale agricultural development programmes, increased extraction for timber and fuel wood, and shifting cultivation. It was envisaged that if this trend continued, the forest cover would drop to an alarming level of 12.5% by the year 2000. The government of Sri Lanka, being mindful of the fast declining forest cover and the increase in fuel and timber needs of the country, commenced an afforestation programme on degraded tea plantations and patna lands. Trees were also planted on the upper catchment of the major rivers as part of the watershed management programme. The forest cover now stands at 27% (**Fig. 3**).

2.   Tobacco Industry: Fuel wood usage and Supply

Cigarette tobacco is cured in barns where heat   is conducted through a set of flue pipes. Hence, it is called "flue curing" of tobacco. Heat is generated in a furnace by burning wood fuel, which is the main source of energy to cure tobacco. Curing of tobacco falls into four stages:

1) ripening at 95^{0}-100^{0}F for 20 hours;
2) colour fixing at 105^{0}-120^{0}F for 12 hours;
3) lamina drying at 125^{0}-140^{0}F for 24 hours; and
4) midrib drying at 145^{0}-160^{0}F for 40 hours.

Studies conducted by the Ceylon Tobacco Company (**CTC**) in the early 1980s revealed that the tobacco industry requires around 100,000 cubic metres of fuel wood per annum to cure tobacco. Although the fuel wood consumption of the tobacco industry was less than 2% of the national consumption, being mindful of the diminishing sources of supply, the CTC embarked on a special programme to reduce usage and ensure fuel wood supply. The programme aimed at the following:

1) fuel-saving devices;
2) alternate fuels; and
3) fuel wood plantations.

Fuel-saving Devices

Initially the CTC was successful introducing a device known as a **warm air furnace**, which reduces usage by 35%. By this device, the hottest part of the furnace (namely the cast iron reducing pipe) is covered to prevent loss of heat and the conserved heat is distributed to the four corners of the barn through additional inflow of air from two inlet ducts situated on either side of the furnance door. The flue system in the barn was subsequently modified into multiflues, which increased the area of radiation and resulted in a saving of 10 %.

While these devices were introduced in the 1980s, conversion of furnaces into the **Venturi system** commenced during 1991. Under this system, the furnace floor is modified into a slanting position from either side leaving only a single gap in the centre, thus reducing the inflow of cold air. Venturi furnaces will bring about a saving of 15%. Installation of warm air furnaces has been completed, while the programme to install multifules and Venturi furnaces is in progress. On completion of these programmes the fuel wood requirement of the tobacco industry will be reduced by 50%. (Table 2).

Table 2. <u>**Fuel wood usage reduction in the tobacco industry**</u>.

	<u>Saving</u>
<u>Saving devices</u>	
Warm air furnace	35%
- Device used to insultate hottest part of the flue system and distribute the conserved heat	
Multiflues	10%
- Increase in area of radiation	
Venturi furnace	15%
- Reduction in excessive inflow of cold air	
Total reduction in fuel wood usage	50%
<u>Alternative fuels</u>	
Coir fibre dust briquettes	
Paddy husk	

Alternative Fuels

Simultaneously the CTC explored the possibility of using alternative fuels for curing of tobacco, and two waste products, coir fibre dust and paddy husk, were identified. The CTC's initial plan was to make fuel briquettes out of coir fibre dust as a substitute for fuel wood. The cost of production of coir fibre briquettes was found to be high in comparison with the price fuel wood on the open market. Hence, the plan to make coir fibre briquettes was not pursued. However, trials conducted in the late 1970s resulted in the successful development of a technique to use paddy husk directly as a source of fuel. Paddy husk is now being used in areas where paddy rice is grown.

Fuel wood Plantations

Although the above-mentioned devices were introduced with a view to reducing fuel wood usage, the CTC's main goal was to make the tobacco industry self-sufficient in its fuel wood requirements. With this in mind, the CTC initiated a programme of afforestation with tobacco farmers as far back as 1976, by supplying them with seedlings of fast-growing trees recommended by the Forest Department for fuel wood production. As of 1991, the extent established by the farmers amounted to 500 acres. In addition, the CTC began its own afforestation blocks on land leased by the government. By the end of 1991, the extent planted by the CTC alone amounted to 1,250 acres. Thus, a total extent of 1,750 acres had been established by the tobacco industry by the end of 1991. The varieties planted are Eucalyptus grandis and Eucalyptus camaldulensis. The CTC's policy is to harvest after eight years and thereafter continue with two coppicing crops (Table 3).

3. Tobacco Industry: Soil Conservation

Any type of seasonal cropping practiced on hilly terrain lends to soil erosion. Recognizing this, the CTC launched a programme of soil conservation with the assistance of the Department of Agriculture in the early 1970s. The progress of this programme was reviewed during the late 1970s and thereafter greater emphasis was placed on permanent methods of soil conservation, such as stone terraces and bench terraces. In 1990, the company initiated a programme of conservation farming by introducing **sloping agricultural land technology (SALT)** to tobacco lands in the hilly country. This technique replaced the mechanical methods of conservation adopted earlier. According to SALT, fast-growing leguminous plants are grown in hedge rows established at determined intervals on the contour. While the hedge rows act as a barrier to prevent soil erosion, the bio mass produced from periodic lopping of the hedge rows enhances soil fertility. Thus, this form of conservation not only prevents erosion, but also helps to recuperate depleted soils.

4. Recycling of Waste Paper

The National Paper Corporation of Sri Lanka uses around 10,600 metric tonnes of waste paper per annum, which represents 45% of their total raw material inputs.

Table 3. CTC fuel wood plantations (acres).

District	1981	1982	1983	1984	1985	1986	1987	1990	1991	TOTAL
Kandy	120	100	50	--	--	--	--	--	--	270
Nuwara Eliya	40	100	--	--	--	--	--	--	--	140
Badulla	40	--	--	--	--	--	--	--	--	40
Matale	--	--	150	70	180	75	175	75	75	800
Total	200 ==	200 ==	200 ==	70 ==	180 ==	75 ==	175 ==	75 ==	75 ==	1,250 ===

Varieties:

Eucalyptus grandis   - Kandy, Nuwara Eliya, and Badulla districts

Eucalyptus camaldulensis   - Matale district

Figure 1.   Agroecological regions of Sri Lanka

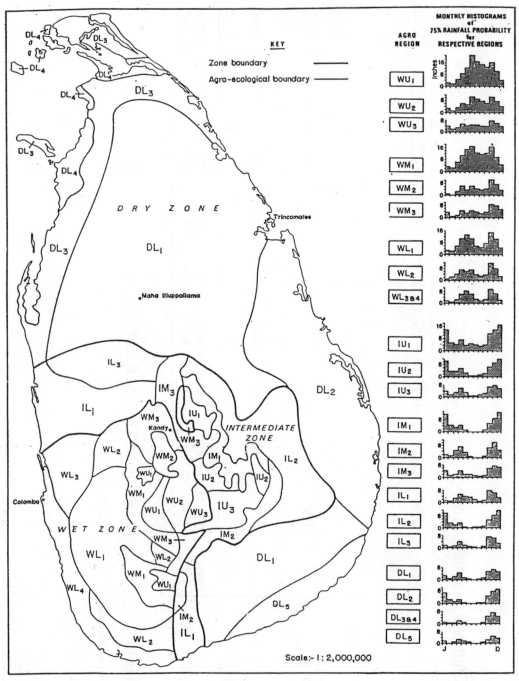

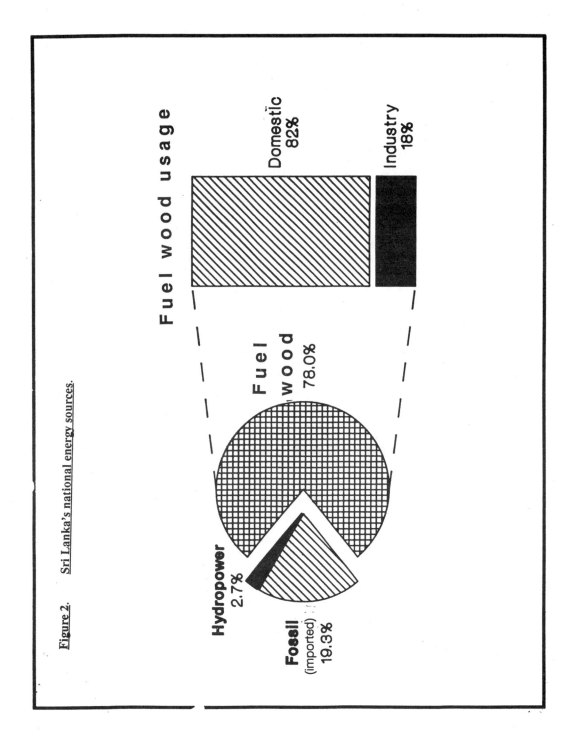

Figure 2. Sri Lanka's national energy sources.

Figure 3. **Land usage in Sri Lanka**

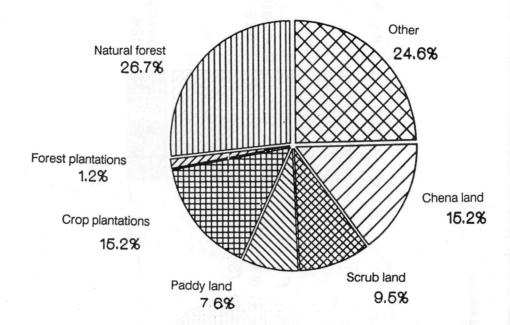

Natural forest
26.7%

Other
24.6%

Forest plantations
1.2%

Chena land
15.2%

Crop plantations
15.2%

Paddy land
7.6%

Scrub land
9.5%

Total land extent: 6,560,000 ha

COUNTRY PAPER: THAILAND

Jitt Kongsangchai
Director
Forest Management
Division
Royal Forest
Department
Bangkok, Thailand

1. Introduction

Thailand covers an area of 513,115 square kilometres, and lies between 5.45^O to 20.30^O north latitude and 97.30^O to 105.45^O east longitude. It is situated on the Indo-China Peninsula in the Southeast corner of Asia with the Shan States and Laos on the north, Laos and Cambodia on the east, the Peninsula of Malaysia on the south, and the Myanmar on the west. A coastline of about 1,930 kilometres is on the Gulf of Thailand in the South China Sea, and the one along the Andaman Sea is about 490 kilometres long. The country's north-south axis is at most 1,650 kilometres and the east-west axis is about 770 kilometres. The narrowest strip is at the Isthmus of Kra Buri and is about 15 kilometres wide.

In Thailand, the forest constitutes a vital natural resource. Besides the economic value of timber and forest products, forests also contribute significantly to maintaining a balanced ecosystem since they help to moderate climatic conditions and to maintain normal seasonal rainfalls, give protection against winds or storms, reduce flood damage by containing the force of flashfloods and resulting soil surface erosion, maintain adequate moisture in topsoil and in the air, are sources of natural fertilizers, and through promotion of infiltration help to augment groundwater resources and to maintain dry season flows in streams. Thus forests are a vital component in a complex ecosystem which act as a buffer for the earth against direct impacts from the sun, wind, and rain.

Forests also have a profound influence on the variety of fauna and flora in the region, and total deforestation would be extremely disruptive to the natural pattern of biota. The climatic conditions would change, resulting in sudden droughts or floods, with accompanying soil erosion in the absence of trees to impede the force of the current. The destruction of a forest's ecosystem would trigger a chain reaction on other ecosystems as well, since all ecosystems are interrelated. Deforestation must therefore be regarded as a potential cause of the ultimate destruction of the overall environmental system on which the human race depends.

2. The Forest Types of Thailand

Being under the influence of monsoonal climate conditions, the vegetation of Thailand is really a humid tropical one, and vast areas are well covered with luxuriant forests. Owing to the composite nature of the topographic factors, climatic factors, edaphic factors, and biotic factors, the forest types of Thailand are considerably varied. The forest types can be classified into two categories: evergreen and deciduous (Samittinand, 1973).

The Evergreen Forest

The evergreen forest is composed of a great proportion of the non-shredding species, and forms about 60% of the total forest area. It can be subdivided into the five following types:

Tropical evergreen forest. This type of forest occurs in and above the wet belt of the area where high rainfall (1,500 mm prevails, and is affected by the monsoon. It is scattered either on the peneplain or in the mountains. This type can be subdivided into three categories as follows:

Tropical rain forest. In the area where contact with the monsoon is direct, the precipitation exceeds 2,500 mm per annum, and nearly the whole area is covered with this type of forest. It generally occurs below 1,000 m in elevation. Under careful study, two zones can be recognized: the lower tropical rain forest and the upper tropical rain forest.

Dry and semi-evergreen forest. This type of forest is scattered all over the area along the depressions on the peneplain, and along the valleys of low hill ranges of about 500 m elevation, or forming galleries along streams and riverlets. The annual precipitations between 1,000-2,000 mm. The forest has three storeys. The upper storey consists of Anisoptera oblongga, A. costata, D. turbinatus, Hopea oderata, H. ferrea, etc. The middle storey is composed of Cratoxylum maingayi, Chaetocarpus castanicarpus, Castanopsis nepheloides, etc., while the lower storey is represented by tree species of smaller structure of the genera Memcylon, Cleistanthus, Aporusa, Alchonia, etc. Palms belonging to the genera Calamus are sparsely present. Bamboos of genera Gigan tochloa, Bambusa, and Dendrocalamus are also sparsely present. Climbers are abundant, belonging to the genera Phanera, Lasibema, Derris, Entada, Strychnos, Sccuridaca, etc.

Hill evergreen forest. This type of forest is confined to upper elevations from 1,000 m upward with high humidity as explained by the moss-clad trees; the precipitation is 1,500 mm annually. It is known as temperate evergreen forest or lower montane forest. The forest has two storeys, dominantly composed of numbers of oaks and chestnut, laurels, magnolias, and rhododendron families, and gymospermous elements are also present, such as Podocarpus, Dacrydium, Cyphalo taxus, Gnetum, and Cycas. The soil is either red granitic, brown-black calcareous, or yellow-brown sandy. Along the valleys the forest is luxriant and rich in species, whereas the summits and ridges offer poor vegetation. The upper storey is represented by Schima vailichii, Cinnamomum sp., Fraxinus excelsa, Dacrydium elatum, Podocarpus imbricatus, Quercus, Lithocarpus, Castanopsis, etc. The lower layer is composed of small trees of medium height and girth such as Gordonia sp.,

Camellia, Acer, Carya, Rhododendron, etc., Bamboos, palms, ferns, sphagnums, and epiphytes are also found in this type.

Pine forest. This type of forest is scattered in small pockets in the north and west continental highlands at elevations of about 600-1,400 m, where poor acid soil occurs. In the Korat Plateau, northeastern part of Thailand, pine forest occurs at lower elevations of about 200-300 m. The soil is either greyish sandy, or brownish gravelly and sometimes lateristic. The annual rainfall is about 1,000-1,500 mm. The composition of the forest consists of only a few pine tree species; Pinus Kesiya and Pinus merkusii are predominant.

Mangrove forest. In Thailand mangrove forest occurs on thick, muddy tidal swamps at the mouths of rivers and along the seacoast in the Gulf of Thailand and on the east and west coasts of the peninsula. The most important trees are mostly of the family Rhizophoraceae such as Rhizophora mucronata and R. apiculata, Bruguiera gymnorrhiza, B. cylindrica, B. parviflora, B. sexangula, Ceriops tagal, etc. Undergrowth trees occur, usually on higher ground not frequently inundated including Acanthus ebracteastus and Acrostichum aureum, Nypa fruticans is usually cultivated, and on the banks Thespesia populnea and Hibiscus tiliaceus are conspicuous with their gaudy yellow flowers.

Swamp forest. Freshwater swamp forest occurs both in central and southern Thailand. In central Thailand the most common species are Hydnocarpus, Xanthophyllnm glaucum, Melaleuca leacadendron, etc.

Beach forest. Beach forests are found along the coasts on sandy beaches. The principal species are Casuarina equisetifolia, Calophyllum inophyllum, etc.

The Deciduous Forest

Deciduous forests are found in all types of conditions and soil on the plains as well as in the hills, particularly along the dry belt of the country where precipitation is low (under 1,000 mm). The climate is more seasonal and the soil is either sandy or gravelly loam and sometimes lateristic. The vegetation of these regions is classified as deciduous formation, where tree species shed their leaves during the dry season. Trees growing in this forest type tend to develop growth or annual rings. The height of predominant trees is comparatively lower (20-25 m) than that of those in evergreen forest. The forest is subject to ground fires during the dry season. The deciduous forest can be subdivided into two types:

Mixed deciduous forest. This forest type is composed of all deciduous species in nearly equal proportions but in certain localities a species may become predominant, such as teak (Tectona grandis linn). Teak forest is open with teak growing on alluvial grounds. In suitable localities teak sometimes forms pure stands. The following are some of the more common trees associated with teak in the mixed deciduous forest: Pterocarpus macrocarpus, Xylia kerrii, Lagerstroemia calycculata, L. tomentosa, Afzelia xylocarpa, Adina cordifolia, Vitex sp. etc. The undergrowth is varied and of characteristic composition. It is chiefly composed of bamboo and many kinds of undershrubs, and in places small palms and cycads appear. Mixed deciduous forests without teak are found scattered in the central and northern portions of the Thai peninsula. The mixed deciduous forests can also be classified into three types based on the terrain and climatic factors:

    1)   most upper mixed deciduous forest;
    2)   dry upper mixed deciduous forest; and
    3)   lower mixed deciduous forest.

    The deciduous dipterocarp forest.   This forest type occupies large stretches of the forest areas in northern, central, and northeastern Thailand. This forest type is found on porous, well-drained soil, which is generally formed by the decomposition of laterite. The soil is either red clay, reddish or pinkish loam, or reddish or white sandy soil, sometimes of poor composition. In northern and central Thailand, deciduous dipterocarp forest occurs both on the level plains and in the hills up to an elevation of about 1,000 m. The general appearance of these forests is open and grassy, often approaching the savannah type of forest; the trees are scattered and as a general rule of medium or small size, both in height and girth.

    The undergrowth consists of long grass and scattered bushes. The forests are burnt over regularly every year; nevertheless, natural regeneration occurs fairly regularly and often luxuriantly. The seedlings are burnt back every year but owing to the accumlation of food reserves in their root stocks, they are able to send up more vigorous and bigger shoots each year. When the shoots are finally free from the danger of fire, the seedlings are able to estabish themselves and become part of the forest crop. The most common trees in these forests are Dipterocarpus tuberculatus, D. obtusifolius, Shorea obtusa, Pentacme siamensis, etc.

    The most important forest type is tropical rainforest, accounting for 43.7% of the total area under forest, and there are significant areas of dry dipterocarp forest (31.18%) and mixed deciduous forest (21.83%). forest area was lost at a rapid rate throughout the 1970s and 1980s by commercial logging and agricultural encroachment. The proportion of the country under forest in 1978 was estimated at 34.15% but by 1989 this had fallen to 27.95%. Less than 15 million ha are now considered to be under forest cover. The types of forest which have suffered the heaviest losses as a result of deforestation are tropical rainforest and dry dipterocarp forest.

3.   **The Forest Area**

Sixty years ago, the almost 513,115 square kilometres which make up the total area of Thailand were covered with dense forests distributed all over the country except part of the great central plain where the forests had long ago retreated due to agricultural activities. The increasing population growth in

Thailand during the last few decades has been at a relatively high rate of 3% per annum. All existing bare ground has been exploited for agricultural purposes. This is the main cause of a very large area of natural forests being encroached upon and occupied by cultivators. Developing countries today are largely dependent on the rational utilization of their national resources for economic growth. Most such countries are becoming aware of the potential of natural resources in their socioeconomic development and of safeguarding the natural environment for posterity. Therefore, many countries are trying to preserve and manage their existing forest resources to the best advantage.

Moreover, foresters have recognized that the forest is an integral part of the ecosystem and its occurrence, structure, and growth are intimately linked with the physical character of the landscape. Management of the forest is actually management of forest land. In addition to essential information on the quality and extent of forest cover, a comprehensive knowledge of the physical characteristics of the land, and past and present and use practices is needed. In order to locate and indicate what, where, and how much of forest resources need to be preserved and managed, remote sensing has been used for surveying and mapping to assess forest resources in Thailand since 1973.

Realizing that forest resources are directly and indirectly beneficial to the economic and social development of Thailand, the Royal Thai Government in the five-year National and Economic Development Plan Phase 4 for the period 1976-1981 set a policy of conserving 40% of the total area of the country (205,246 sq. km) as a forest area. However, due to political, economic, and social pressures, the forest condition has consistently deteriorated. This means the set goal has not been reached. If we do nothing, within the next two decades Thailand's forestry resources will be so reduced as to cease making any economic contribution to the country. Therefore, there is an urgent need to use modern techniques of remote sensing for surveying and mapping forest resources in order to solve all of these problems. The results from using the remote sensing techniques for forest assessment since 1961 are shown in Table 1.

Table 1. **Forest assessment using remote sensing techniques.**

Year	Forest area (sq. km)	%	Remarks
1961	273,628	53.33	By aerial photos
1973	221,707	43.21	By satellite data
1976	198,417	38.67	"
1978	175,224	34.15	"
1982	156,600	30.52	"
1985	149,053	29.05	"
1988	143,803	28.03	"
1989	143,417	27.95	"

The change in forest area in Thailand in the past 28 years (1961-1989) was approximately 129,000 sq. km or 4,800 sq. km per annum. Forest area by type as of 1988 is shown in Table 2.

Table 2. **Type of forest area in Thailand in 1988.**

Forest types	Area (sq. km)	%
Tropical evergreen	62,850	43.70
Mixed deciduous	44,837	31.18
Dry dipterocarp	31,390	21.83
Mangrove	1,963	1.37
Pine	1,983	1.38
Scrub	775	0.54
Total	143,803	100.00

4.   Forest Management in Thailand

In Thailand, all natural forests and forest lands are state owned. The government body responsible for the forests is the Royal Forest Department (RFD) which is part of the Ministry of Agriculture and Cooperatives. Management and control of activities in forests or on forest lands are decentralized to 21 Regional Forest Offices and 72 Provincial Forest Offices. The RFD's primary objective is to maximize the benefits of the forests for the greatest number of people, both in terms of forest production and protection of the resources. The government has faced substantial problems in realizing these objectives.

The National Forest Reserve Act was passed in 1964. This empowered the RFD to designate forest reserves, which were then gazetted by Royal Decree. Forest reserve status applied to permanent forest as defined by the National Land Classification Committee. In 1989, the RFD declared over 1,200 separate forest areas as forest reserves, covering a total area of 23.2 million ha. Within the forest reserves, there are 226 areas of forest preserved under other nature conservation designations, amounting to 6.5 million ha. However, the actual forest cover of the whole country in 1989 was less than 15 million ha.

Prior to 1988, selective logging was permitted in forest reserves under a logging concessionary system. However, in response to great public concern over the rate of forest loss and widespread occurrence of illegal logging, the government instituted a nationwide ban on logging in forest reserves and other concession areas. All logging concessions except those applying to mangrove forest are now in suspension until the ban is lifted.

The framework for forest management is set out in the National Forestry Policy. The policy has set the minimum requirement for forest coverage at 40% of the country's land surface. Forest areas are divided into two parts for the purpose of the policy: conservation forest (composed of first class watersheds, national parks, wildlife sanctuaries, non hunting areas, reserved parks, arborea, botanical gardens, and reserved areas for specific studies); and commercial forest or productive forest (composed of the remaining forest reserves, plantation areas, community forests, private true farms, and timber concession areas). The total target area for forest cover is divided between these two forest types. Under the Seventh Plan, the proportion of conservation forest is set at 25% and commercial forest at 15%. This demonstrates the government's commitment to stopping the process of forest loss and sets an ambitious target for reafforestation, to raise the forest area from 28% overall to 40%. Moreover, it emphasizes the need for conservation forest to form the majority of the forest cover and seeks to build on past success in establishing wildlife sanctuaries and national park areas.

Reafforestation

The principle of reafforestation has been accepted in Thailand for almost a century. Most early plantations, however, were designed to increase timber production rather than to achieve any environmental objective. Since 1967, plantation projects have been established by two government agencies and one state enterprise. Only in the last decade has the private sector become involved in the reafforestation programme. However, in the past the government has not been successful in achieving its target areas of forest set out in the National Forestry Policy. This is partly the result of a lack of financial resources and political support for the RFD to enforce forestry policy, and also because of the strength of the social and economic factors encouraging the rural poor to clear forest land for agriculture.

Apart from natural forests, there are forest areas planted by both the government and the private sector. Thailand's RFD reports that the total of this area in 1989 covered 6,968.94 square kilometres. Planted areas in the north, northeast, centre and east, and the south were 3,835.70, 1,338.97, 1,088.13, and 706.14 square kilometres respectively. The major tree species planted were <u>Tectona grandis</u>, <u>Pinus kesiya</u>, <u>Eucalyptus</u>, <u>camaldulensis</u>, <u>E. deglupta</u>, <u>Acacia auriculaeformis</u>, <u>A. mangium</u>, <u>Casuarian equisetifolia</u>, <u>Hevea brasiliensis</u>, <u>Rhizophora apiculata</u>, etc. Annual reafforestation in Thailand has been shown in Table 3.

Table 3. **Reafforestation projects in Thailand (square kilometres).**

Planting type	Up to 1980	1985	1986	1987	1988	1989	Total
Afforestation	1,973.95	106.40	102.40	100.53	99.30	85.55	2,468.13
Watershed rehabilitation	960.93	98.70	93.12	87.36	83.47	78.14	1,401.72
Reafforestation	893.19	75.87	75.78	75.68	78.80	38.59	1,207.24
Concessionaires' reafforestation	1,076.01	100.18	81.99	103.89	93.60	12.03	1,467.70
Planted by Forest Industry Organization	249.26	9.30	6.45	5.14	4.21	3.42	277.78
Planted according to ministry regulation	-	-	0.75	23.69	26.92	32.01	83.78
Total	5,153.34	390.45	360.39	399.41	385.61	279.74	6,968.94

In response to the concern over Thailand's diminishing forests, the government has initiated two programmes to try to increase forest cover in Thailand: commercial plantation of fast-growing trees and a programme of community, or social, forestry. Under the commercial plantation programme, the government provides incentives to the private sector to plant fast-growing tree species on encroached and degraded forest land. The incentives include renting encroached land to private companies at competitive rates and granting promotional privileges to participating companies. The government has promoted fast-growing species such as eucalyptus and acacia so that forest cover can be reestablished quickly and greater levels of rural employment can be created. It is envisaged that some farmers may even plant their own lands with fast-growing trees, for which a ready market exists in the woodchip and pulp and paper industry.

The community forestry programme has been formulated in recognition of the fact that poverty is a major underlying cause of deforestation and that any reafforestation project must tackle the issue of poverty if it is to succeed. Community forestry involves members of the community in all aspects of the decision-making related to that enterprise. The premise is that local people know best how to protect their own forest resources and have the greatest incentive to do so. The essential elements of community forestry are:

1) generation of income and stable employment for local people;

2) sustainable production of forest products such as wood for fuel and construction, fodder, and food for the community;

3) control of local ecological land degradation and maintenance of land productivity; and

4) strengthening of rural community institutions.

Community forestry has been practised in some parts of Thailand for many years. The community-initiated Muang Fai system of protecting local watersheds in the forest which were threatening water supply to the rice fields is a good example from northern Thailand. A number of schemes have been established by the RFD and the Forest Industry Organization in the past, which comply with some or all of the principles of community forestry.

Nature Conservation and Protected Areas

As Thailand's once abundant natural resources become scarce, conservation of natural resources has received growing attention. Natural resource conservation, however, can only be fruitful if serious control is enforced by government agencies. The various types of natural conservation and protected areas established in Thailand are shown in Table 4.

Table 4. **Conservation and protected areas in Thailand**.

Area	Number	1991 Area (km^2)
National Park	66	34,503.49
Forest Park	35	924.78
Wildlife Conservation area	33	25,860.46
No-hunting area	43	3,470.55
Wildlife park	2	24.56
Botanical garden	5	14.00
Arboretum	42	28.70
Total	266	64,826.54

5. Conclusions

Forests are one of the few renewable natural resources. With careful planning and good reafforestation programmes we can harvest timber indefinitely. Forest land use should be planned in conjunction with the master land use plan. Using remote sensing application for forest land use planning it was found that within the past 16 years the forest area in Thailand has been depleted and deteriorated at the rate of 1.0% or about 5,194 square kilometres annually. The results show that the total forest areas of Thailand is lower than the Nation Forest Policy target to maintain 40% of the total area of the country as forest area. This indicates that Thailand's forest has now reached a crisis stage. Such deterioration of forest is due to population increase and political, economic, and social pressures. All of these pressures result in the forest condition in Thailand becoming worse and cause many problems to be solved.

However, we can plant today for the future. To achieve the desired target of 40% it is necessary to expand the forest plantation area by establishing forest plantation programmes. Moreover, forest resources must be routinely monitored using remote sensing as a tool in order to determine the existing forest area as an input to optimum forest land use planning. The technical cooperation projects between the RFD and other countries or international organizations should be strengthened in order to solve the problems of tropical rainforest deterioration and the greenhouse effect that will increase the temperature of the earth in the future.

III. Report of the Study Meeting

and

New Trends in Environmental Management

REPORT OF THE STUDY MEETING

Background

Tropical forest development projects have been one of the most important and topical issues in global development and environmental protection. While tropical forests are a major source of oxygen, they are also important sources of economic products and are subject to population pressure on the part of both producing and consuming countries. Hence we have to find some way to establish cooperation between producers and consumers that will benefit both and still be environmentally sound. This is where the concept of sustainable development comes into play. The tropical forests are very important from the view point of environment, particularly because external diseconomies such as soil erosion or losses of bioresources and wildlife are difficult to calculate.

Yet there are powerful pressures from producing countries to cut down trees for economic gain. If these trees are to be preserved for all mankind, the cost of preserving these forests must be shared by developed countries. As one of the world's major consumer of forestry resources, Japan should pay serious attention to maintaining the balance between development and the environment. In developed countries, trees are cut down for industrial use but not depleted so quickly due to reforestation efforts. But developing countries have not yet developed the ability to manage their forestry resources. Transfer of technology has therefore become extremely important to alleviate the problem of deforestation. Japan has developed technologies and accumulated valuable experience needed to reconcile development with environmental preservation.

Against the above background, the APO organized a study meeting on New Trends in Environmental Management to share, among others, the Japanese experience and technologies on forestry resources development. The meeting was conducted in Tokyo and Nikko from 3rd to 7th February 1992 and was attended by 11 participants from 9 member countries, together with resource persons from Indonesia and France as well as the host country. The programme of the meeting and a list of participants are attached as Appendices I and II, respectively.

Summary of Presentations and Discussions

1. Japanese Experience in Pollution Prevention and Environmental Protection: An Overview

The outbreak of industrial pollution typically represented by itai-itai disease and Minamata disease prompted the Japanese government to control pollution through legal measures. In the early 1970s, a series of laws and regulations were enforced which have proved to be highly effective in combating pollution in Japan. Gradually, however, pollution problems have been succeeded by environmental issues that could cause disastrous effects on a global scale. These include, for example, greenhouse effects, the destruction of the ozone layer, the disappearance of tropical forests, acid rain, and treatment of hazardous waste. Economic activity is undertaken based on decisions made by a company (within the framework of domestic laws and regulations), but there are always possibilities that such decisions will affect other companies or countries. For example, Japan has imported timber resources through the market mechanism. But the price paid does not necessarily reflect the cost of reforestation or damage from soil runoff after tree-cutting. In this context, it is necessary to define the "externality of environmental deterioration" caused by industrialization and urbanization.

From the Meiji Restoration in 1868 until the era of high economic growth in the late 1970s, the Japanese government, industries, and the society in general have paid no particular attention to environmental degradation. Firms have only sought to generate maximum profits from industrial activities. In case of the Ashio mine, one of the oldest mining operations in Japan, crop irrigation water was polluted by effluents contaminated with copper, which forced victims from the land they had farmed for many generations. Still no attempts were made by the government or the firm to stop the discharge of effluents without proper treatment.

Around the end of the 1960s there was an outbreak of a series of serious pollution problems which occurred one after another: Minamata disease, itai-itai disease, and Yokkaichi asthma. These incidents sounded alarm bells throughout Japan, and local residents and governments started to undertake antipollution campaigns. In response to awakened public opinion, the Japanese government set environmental standards and related effluent standards to protect the health of the people and the living environment. These regulations directed firms to observe effluent standards by installing treatment facilities. At the same time strict regulations motivated them to renovate production processes so as to reduce the discharge of effluents. In this process they came to realize that investment for environmental protection results in higher productivity in view of the reduce use of energy and resources. The fact that new investment in pollution prevention generates profit has had ripple effects among firms, which contributed eventually to overall environmental improvement at the macro level in spite of the rapid economic growth during the 1960s and 1970s.

The efforts toward pollution prevention seemed to draw a conclusion in the debate of whether the relationship between economic growth and environmental protection constitutes a trade-off relationship or not. The Japanese experience of successful control over water and air pollution during the period of high economic growth seems to support this conclusion. Further it was about 20 years ago when the concept of "PPP" was introduced. Its conventional meaning was "polluters pay the principle." This meant that all externality generated by firms should be internalized at the cost of the pollutant. In fact, the Ashio reforestation project showed that the recovery of natural forest cost much more than pollution prevention. Once natural conditions were degraded, reinstatement to the previous condition has been found to be extremely costly and time consuming. Thus pollution prevention brings economic benefits. Nowadays, "PPP" is interpreted as "pollution prevention pays".

Pollution prevention is also an economic activity and hence relevant regulations and laws should be based upon economic considerations. Many economists have argued that indirect regulation through the price mechanism was better than direct regulation. The main proponent of this argument was A.C. Pigou, who treated environmental problems as "technological externalities," and he recommended tax and subsidy policies to obtain a socially optimal output. More recently, J.H. Dales of Toronto University has proposed the idea of a "discharge claim exchange system" using the market mechanism. Basically, however, economic policies and regulation have innate defects in coping with environmental issues since they cannot be simply solved by the Pigouvian externality approach. In classical environmental problems, the causer and sufferer can be clearly identified, and further the problems do not go beyond national borders. Nowadays environmental issues must be considered in global terms, however. Once CO_2 has accumulate sufficiently to affect the weather, it is impossible to reinstate good weather. Moreover, the adverse effects on human will appear only a generation or longer. If the optimal discharge level of CO_2 must be decided now, policies and regulations should also reflect the interest of future generations, taking into account changes in economic conditions which may influence the future global environment. A new theory on environmental economics is still in the embryonic stage and has not yet been established as full-fledged theory due partly to the paucity of data and empirical evidence. Establishment of a new economic theory is now called for since policies and regulations regarding environmental control cannot be designed unless they take into account economic activities across national borders and beyond the present generation.

2.   **Forest Preservation and Reforestation**

A forest itself prevents floods by facilitating the permeation of rainwater into the soil, and serves as a natural reservoir. Furthermore, forests serve not only to prevent landslides and soil erosion but also to alleviate the damage caused by blown-off sands, wind, snow, and fog, and to check the spread of fire. Other merits of forests are their abilities to purify air, absorb noise, and provide us with recreation sites. Thus forests are indispensable resources not only for soil conservation but also for our quality of life. In developing countries, forests are part of life and a main support of economic activities (tropical forest covers as much as 50% of the total land area of Southeast Asia). However, unplanned harvesting results in various externalities as mentioned above. What must be prompted and perhaps enforced and assisted by developed countries are methods to:

1)   rehabilitate critical forest areas (due to forest fire, inappropriate shifting of cultivation, etc);

2)   replant logged forest with the most appropriate tree species;

3)   promote forest industries that can generate higher value added; and

4)   promote nonwood forest products and wood plantation.

Two common terms emerged from the discussion of the meeting: sustainability and community participation. These suggest the importance of the involvement of local people in environmental preservation projects so that they will recognize the need to maintain a balance between resource development and environmental protection. In this connection, the commitment and achievements by Nissho Iwai Trading Company with its afforestation projects in Papua New Guinea continuing over nearly 20 years should be emulated.

Environmental protection in relation to forest resources is particularly important in the light of the magnitude of negative impacts upon the region. In Japan the river and water pollution caused by industrial effluents has been eliminated to a large extent, but the damages to forest resources cannot be recovered so quickly. Reforestation projects in the Ashio region have been continuing for more than 40 years. The cost of reforestation including soil erosion protection through the construction of check dams costs approximately US$1 million per hectare, whereas that of aerial seeding costs about US$20,000 to 50,000 per hectare. Reforestation consists of various processes including the stabilization of slopes along rivers and streams by constructing check dams, concrete and log wall works, grass and tree seeding, etc. In case of manual operation, the process undertaken in the first year is to construct retaining walls and hillside works to stabilize the surface of the slope. Grass planting using the horizontal grassing method and vegetation sacks is implemented during the second year. In the third year, the seeding of pines, birch, and pseudoacasia is carried out. Clethra is dominant in aerial seeding.

A Japanese forestry engineer developed a new technology for reforestation, which was a compact soil cake mixed with soil, seeds, and fertilizer. The cake, placed on hillsides proved to be very effective through Japan. This technology has been further improved in the form a "vegetation sack" that contains all necessary ingredients as in the soil cake. The productivity of reforestation was increased due to these technological innovations until aerial seeding was introduced. Currently, both manual and aerial seeding methods are used based upon the technical evaluation of the project.

External diseconomies caused by reforestation and soil erosion prevention in the Ashio region were not internalized by the company responsible for undertaking copper mining operations. However, the firm was obliged to treat drainage from the old mine by installing a water treatment plant using the method of chemical coagulation. Instead the responsibility has been assumed by both local and central governments. Some of the problems they have faced in reforestation are:

1)   Due to a shortage of workers as a general labour problem in Japan, it is extremely difficult to obtain workers for the reforestation project. Hence the average age of workers is becoming older.

2)   Aerial seeding is highly effective from the economic viewpoint, but this is not fool-proof since check dams and retaining walls are also needed.

3)   To protect nursling trees from wild animals, they have to be sealed in net sacks, which raises the cost of reforestation.

4)   It takes time to recover vegetation in the Ashio valley.

There is a linkage between forest management and paper manufacturing. In particular, the consumption of forestry products needs to be studied, including ways to consume products and waste products to encourage and enforce recycling in producing countries.

Japan ranks second in the per-capita production of paper, third in consumption, and first in waste paper recovery and utilization in the world. However, almost all waste paper recovered is newsprint and corrugated cardboard, which are recycled into newsprint, corrugated cardboard, and toilet tissue. The recycling system of such paper is affected by the market price of used paper, and in fact revenues derived from the sale of waste paper are less than the cost required for collection and transportation. Given different circumstances, government subsidies may be required for recycling from the viewpoint of conserving global resources.

Meanwhile paper for computer and copying machines, generally called PPC (plain paper copier) was recycled less until 1989 in Japan. Recently, however, a new recycled PPC paper that is almost as same as the conventional PPC paper in quality and cost has been developed. It is characterized by 70% waste content, alkalinity, high brightness, and high machine reliability. These features result from deinked pulp (DIP) manufacturing technology. Now almost all large paper manufacturing firms have introduced DIP equipment so that the share of the recycled PPC paper in all PPC paper sales has increased from 1% to 20%. Not surprisingly, recycling systems designed and implemented on the basis of a single [e.g., Fuji Xerox's Protection and Help Operations of Environment Nature Initiated by Xerox (PHOENIX) Activity] or a group of companies (Office Town Association System sponsored mainly by the Tokyo Electric Power Company with the participation of more than 50 large firms) has been working more efficiently than those organized on a community basis.

In order to preserve forests, the forest protection system was established in 1987 based upon the Forest Law. Forests are registered in accordance with the law and are placed under certain restrictions in their management. In Japan there are three restrictions imposed on the management of protected forests: restrictions on felling of standing trees, the conversion of topography, and the obligation of planting. Felling of standing trees is to be undertaken for the felling system (clear cutting, selected cutting, or cutting prohibition), and the annual cutting area and the modification of topography require the permission of the prefectural governor. After felling, it is required to commence planing within two years. If the protected forest comes under private property (and this means it can be used for public service only), special favours are granted by the government including tax exemption or reduction, financing, reforestation subsidy, reforestation with the public entity on a profit-sharing basis, etc. At present the protected forests have been classified into 16 types depending upon their purpose (water conservation and erosion control are the two major purposes).

In Japan, projects for prevention of sediment-related disasters are undertaken by the Ministry of Construction and the Forestry Agency, under the Ministry of Agriculture, Forestry and Fisheries. The works undertaken by the former ministry are called <u>sabo</u> works (sediment disaster preventive works) with a view to protecting human lives and property in basins. <u>Sabo</u> works include sediment control projects, construction to prevent landslides, and additional construction to prevent steep slope failures. Those undertaken by the latter agency are called forest conservation works. They include forest conservation in mountainous areas, in particular, mountainous disaster hazard areas, development of disaster prevention forests, protection of forest maintenance works, and landslide prevention.

It is important to invite policy makers and decision makers to this kind of forum so that they can appreciate the importance of land conservation, forestry development, erosion control, and reforestation in the overall framework of environmental protection at the macro level. Further involvement of the private sector, nongovernmental organizations, and women should also be considered for future meetings.

It may be useful to develop functional linkages among forest areas, farmland, and marine areas so that the optimal use of forestry resources and afforestation will eventually affect other natural resources available on lowlands and by the sea.

NEW TRENDS IN ENVIRONMENTAL MANAGEMENT

Arata Ichikawa
Department of Urban
Engineering
University of Tokyo
Tokyo, Japan

1. Role of the Asian Productivity Organization in Solving Environmental Problems in Asia

Since its inception, the Asian Productivity Organization (APO) has had as its main goal the diffusion of systems and techniques to enhance productivity, especially in industry. As an example, the total quality control (TQC) system has been proven one of the most successful methods to improve productivity and has spread throughout Asia and the world.

Because rapid industrialization has created deleterious environmental effects, however, the APO has also organized seminars on environmental issues since 1980. In that year, the APO held a study meeting on "Managing Industrial and Agricultural Wastes" in Tokyo, followed in 1982 by a second meeting on "Treatment and Disposal of Hazardous Wastes from Industry," also in Tokyo, attended by Japanese government-related corporations and representatives of private industry. The "Eutrophication of Lakes" was the topic of the third Tokyo meeting in 1984, at which participants were introduced to a computer software system that simulated models of lake eutrophication and allowed testing of possible countermeasures for this serious problem.

Organizations concerned with governmental development assistance projects evaluated such APO seminars highly, including the German government. Therefore the fourth environmental meeting was held in Wiesbaden in 1986 after the EC issued its Directive on Hazardous Materials. Two representatives of the Hessen State Green Party were invited to discuss public participation in environmental projects, and representatives of the German Ministry of Environment and government-related corporations were also in attendance. Subsequently, a fifth meeting, cosponsored by the German government, was held in New Delhi and Bombay, India, in 1988. German experts explained the EC Directive on Hazardous Materials, and participants discussed the management of hazardous materials from the points of view of environmental and labour safety.

After the Indian meeting, the APO has continued to support investigations of environmental problems and issues. Conditions have changed drastically in the ensuing years since 1988, and the most urgent problems currently acid rain, deterioration of tropical rainforests, and global warming. Thus the APO decided to host a study meeting on various aspects of forest management in early 1992.

2. Background to Asia-Pacific Regional Environmental Problems

The Asia-Pacific region has been developing rapidly and undergoing concomitant structural change. Along with such rapid development, environmental problems have become serious, especially in the newly industrializing economies (NIEs). Such problems include deterioration of living environments due to increasing concentration of populations in urban areas and environmental pollution of the earth, air, and water. In parallel with increased awareness of regional problems, there has been an increased awareness of global environmental issues such as the greenhouse effect and depletion of the ozone layer.

Japan has a definite role to play in the process of achieving sustainable development in the Asia-Pacific region. It has had to struggle to solve serious environmental problems that occurred during the course of economic development. Sharing those experiences, especially with developing countries, will not only deepen regional understanding of the problems, but lead to common measures taken against them.

There are complex interrelationships between the four factors of population, resources, development, and the environment, and thus the study of environmental problems must cover multiple aspects. Industrialization has led to economic development and increasing urbanization, along with trade and direct investment from developed countries. This has unfortunately been accompanied by deterioration of soil and increasing desertification in China, Australia, the USA, and Africa; by a decrease in the area covered by tropical rainforest in Thailand, the Philippines, Malaysia, and Papua New Guinea; by reduced fishing resources in the North and South Pacific; and by an increase in the number of animals facing extinction.

Developed and developing countries tend to view environmental issues differently. Developed countries take such issues extremely seriously and feel that finding solutions is an urgent task. Developing countries, on the other hand, feel that economic growth in the developed countries was achieved at the expense of their resources and that it is unacceptable now to have to shoulder the responsibility of paying for the short-sightedness of the past at the expense of their own economic development. Global environmental problems, however, require global measures taken by a world community bound together in a common destiny.

Rapid industrialization and urbanization have contributed to polarization between urban and nonurban populations as well as to widespread environmental pollution. While cities in developed countries have the benefit of vast amounts of investment in their infrastructure, urbanization in developing countries has not proceeded in a systematic manner. This has resulted in insufficient infrastructure facilities for rapidly increasing populations. One important example of this lack of infrastructure is the water supply in many cities. Some urban districts are not provided with either waterworks or sewerage facilities. Citizens living in those districts depend on water vendors, who often take water from public hydrants or common wells. Due to the lack of systems for securing sanitary, safe water for drinking, cooking, and washing, infectious disease can be spread more easily. Air pollution is another serious environmental problem accompanying urbanization which affects the health of all.

Differences in people's perceptions of the environment make problems of the living environment more complex, and consequently make finding solutions more difficult. Differences occur not only between developed and developing countries, but also between the rich and poor within a single country. Thus care must be taken when devising solutions to environmental problems and in establishing infrastructure facilities to serve the needs of all sectors of society.

3.   Japanese Experience in Environmental Problems

During the period of high economic growth, government expenditures focused on improving industry-related social capital. The delay in spending to enhance social facilities related to the living environment resulted in widespread industrial pollution, which gradually became more serious in spite of the growing awareness of the dangers of pollution by citizens, businesses, and the government after 1965. When pollution problems culminated in outbreaks of Minamata disease, itai-itai disease, and Yokkaichi asthma, the national industrial pollution policy began to change. Reasons for Japan's unawareness of environmental problems during the 1960s and 1970s include   the following:

1)   It was thought that environmental measures would result in additional expenditures while industry was struggling to reduce costs, and thus were not important.

2)   Since enterprises had little experience in environmental protection at the time, conservation and protection know-how was lacking.

3)   Enterprises did not have the necessary profit margin to invest in environmental protection measures.

Campaigns by residents and efforts made by local public bodies played a very important role in increasing awareness of environmental problems. By the early 1970s, the government had put legislation concerning environmental pollution into effect and industry was obliged to comply with the series of regulations enacted. This brought about a drastic increase in the amount of investment in equipment to prevent industrial pollution, amounting to ¥964.5 billion in 1975, or 17.7% of total private plant investment. The cost of investment related to the prevention of industrial pollution for sustainable development was extremely high in those early years and created a heavy burden for enterprises. Later it declined, and has remained at around ¥300 billion in the past few years.

It should be pointed out, however, that investment in environmental conservation has had positive effects on the national economy. Although prices increased due to high investment, income also increased due to higher demand. Investment in industrial pollution control measures contributed to an increase in GNP in real terms of 0.9%. Rough calculations show that from the macroeconomic viewpoint the effect of the investment was not negative during the past stage of rapid economic growth. Other positive effects can also be pointed out, such as those on the automobile industry. Regulations concerning automobile exhaust emission led to competition between manufacturers in technological development which resulted in better overall efficiency of automobiles.

In 1970 Japan imposed strict regulations on the discharge of hazardous substances, partly as a result of the outbreak of Minamata disease. Many developing countries, however, have judged it unnecessary to introduce such strict controls due to the smaller scale of their industrialization to date. Less stringent regulations are sometimes even used as an inducement to attract multinational enterprises to locate factories in developing countries. "Catch-up" industrialization has thus created serious environmental problems, and countermeasures have come too late in some cases.

Recently many countries have been endeavouring to organize structures and systems to allow the imposition of regulations similar to those in place in Japan. In some cases, substantial improvement has yet to be seen, since laws and regulations are difficult to apply to existing facilities and "exceptional rules" are applied under pressure from high government officials or influential enterprises. Efforts made by the government departments in charge of environmental protection and by local community residents, however, have resulted in strengthening of regulations governing existing as well as new factories.

Because the location of a manufacturing plant of an enterprise from a developed country or of a multinational enterprise in a developing country is considered an effective method of speeding up the development process, it has been requested that "enterprises of developed countries or multinational enterprises should take the same environmental measures regardless of location." When this is practised across the board, no benefits will accrue from "external diseconomy by utilizing differences of regulations." This will also have a positive effect on the environments of developing countries.

Technology for preventing environmental pollution is sometimes not available or its cost is thought to be prohibitive. Developing countries must also consider global environmental problems in addition to national ones brought about by increased energy demand and the amount of CO_2 emitted following industrialization and improved living standards. These increase the "cost of industrialization" through "negative investment" that cannot be afforded in some instances. Developed countries must give assistance in the form of technology, know-how, and adherence to pollution control measures, especially when setting up plants in developing countries. Simultaneously, the generated unit of contamination should be reduced as industrialization proceeds. To achieve this, measures similar to those taken by Japan should be enforced to save resources and energy.

4.   Global Environmental Problems

Over the past decade serious environmental accidents have occurred, especially in Europe and the USA. The explosion at the Union Carbide plant in Bhopal, India (1984), the discovery of the hole in the ozone layer (1985), the accident at the atomic power generation plant in Chernobyl, USSR (1986), which contaminated the Rhine River, the deaths of vast numbers of seals due to North Sea pollution (1988), and the crude oil spill from the Exxon Valdez off the coast of Alaska (1989) are only a few examples. Such accidents that affect the environments of neighbouring countries underline the need for international cooperation in environmental protection. Some international organizations, notably the UN's UNEP, have played important roles in achieving the needed cooperation. International congresses and conferences, including the Montreal Conference on Fluon Gases (1987), the Basel Treatise on Transborder Transport of Hazardous Wastes (1989), and the Arche Summit in France (1989), which was meant to be a conference on economic issues but changed its agenda to reflect the importance of environmental protection, are examples.

Such efforts have led to some improvement in global environmental problems. Technology and hardware cannot solve all such problems alone, however. Further international cooperation is necessary, and such meeting as this one hosted by the APO represent ongoing efforts to exchange information and contribute to technology transfer between countries.

5.   Solid Waste Management: A Topic for the Future

In the past, waste generated from human activities was disposed of through the natural purifying capacity of the water, land, and air. With increasing population and wider-ranging human activities, however, this has become impossible and the amount of waste generated has increased rapidly. While composting, incinerating, and recycling technologies have been developed, there are limits to their application. Budget constraints and the time needed to construct such facilities are but two of the limitations.

For a long period in the USA, hazardous waste was dumped on government-owned, unoccupied land. Currently, the US Environmental Protection Agency estimates that 30,000 dumping sites are dangerous and must be reclaimed in future at great expense. In Japan, the central government has removed the bottom sludge in Minamata Bay to prevent the future recurrence of Minamata disease. Local governments have replaced contaminated soil. Both have been achieved at enormous cost, and the Japanese have learned from the cost of reclaiming polluted land and sea is far higher than the cost of prevention, treatment, and recycling.

To dispose of waste, or to recycle, most effectively, it must be separated into categories. In some Japanese local government areas, inhabitants must separate their waste into four to six types: combustible, noncombustible, plastic, metal, glass, and hazardous (including dry-cell batteries). Old newspapers, cardboard, and office paper are collected separately for recycling, although this recycling system is heavily influenced by the market price for used paper, which tends to fluctuate.

The current Japanese recycling system is thus imperfect. It is hoped that it can be rebuilt with a global perspective on solid waste management.

6.   **Forestry Resources**

Because forestry resources are used for a variety of purposes, ensuring their sustainable development requires first determining the total demand. Reducing the total demand can be achieved through recycling, replanting, and substitution, among other methods. Japan's forests have deteriorated and been greatly reduced in size, meaning that flooding has become a greater danger than in the past. Air pollution has also contributed to reduction in the forest cover. Another report is presented at this meeting on the conditions of Japanese forests and the effects Japanese imports of timber from Indonesia, the Philippines, and Malaysia has had on those countries. After a technical site visit to Nikko, the site of a forest being reclaimed, other possibilities for reafforestation were discussed.

IV. Appendices

List of Participants and Resource Persons

<u>Participants</u>

China, Rep. of

Mr. Kuo-Ching Chie
Senior Specialist (Administrative Forester)
Council of Agriculture, Executive Yuan
37 Nan-Hai Road, Taipei 100
Taiwan

Hong Kong

Mr. Man-kwong Cheung
Senior Conservation Officer
Agriculture and Fisheries Department
Hong Kong Government
14/F, Canton Road Government Offices
393 Canton Road, Kowloon

Mr. Bernard Ian Dubin
Senior Environmental Protection Officer
Environmental Protection Department
Hong Kong Government
EPD 27th Floor, Southorn Centre
130 Hennessy Road, Wanchai

India

Mr. R.V. Krishnan
Principal Secretary
Department of Energy
Forests Environment, Science and Technology
Government of Andhra Pradesh Hyderabad
Hyderabad

Indonesia

Prof. Dr. Affendi Anwar
Professor in Resource Economics and
    Chairman on Regional and
        Rural Development Studies Program
Graduate School
Bogor Agricultural University
Jalan Raya Pajajaran, Bogor

Korea, Rep. of

Mr. Young-Kyoon Yoon
Assistant Director
Forestry Administration
Cheongyangri-1, Dongdaemungu
Seoul

Pakistan

Mr. Mohammad Irfan Kasi
Managing Director
Ziarat Valley Development Authority
Government of Balochistan Quetta
c/o Planning and Development
Government of Balochistan Quetta
Quetta

Philippines	Mr. Joseph M. Alabanza

Philippines   Mr. Joseph M. Alabanza
Regional Executive Director
National Economic and Development Authority
   Cordillera Administrative Region (NEDA-CAR)
Botanical Garden, Leonard Wood Road
Baguio City

                        Mr. Leonardo D. Angeles
Executive Director & Secretary, Board of Directors
Philippine Wood Products Association
3rd Flr. LTA Bldg., 118 Perea Street
Legaspi Village, Makati
Metro Manila

Sri Lanka   Mr. Arumadura Victor Nihal Silva
Leaf Manager
Ceylon Tobacco Company Limited
Post Box No.: 62
175 Paranagantota Road Mawilmada
Kandy

Thailand   Mr. Jitt Kongsangchai
Director of Forest Management Division
Royal Forest Department
Bangkok 10900

Resource Persons

                        Prof. Arata Ichikawa
Department of Urban Engineering
Faculty of Engineering
University of Tokyo
3-1, Hongo, 7-chome, Bunkyo-ku
Tokyo 113, Japan

                        Dr. Takehiko Ohta
Forest Hydrology & Erosion Control Engineering
Department of Forestry
University of Tokyo
1-1-1 Yayoi, Bunkyo-ku
Tokyo 113, Japan

                        Dr. Jean-Marie Fritsch
Senior Hydrologist
French Institute for Scientific Research
   and Cooperation for Development (ORSTOM)
Centre de Montpellier, 2051
Avenue du val de Montferrand - B.P. 5045
France

Prof. Ryuichiro Matsubara
College of Arts and Sciences
University of Tokyo
3-8-1 Komaba, Meguro-ku
Tokyo 153, Japan

Mr. Seiji Mori
Project Coordinator
Afforestation Project Promotion Office
New Business
Nissho Iwai Corporation
4-5, Akasaka 2-chome, Minato-ku
Tokyo 107, Japan

Mr. Taiji Ohashi
Director
Products Support Unit
Fuji Xerox Office Supply Co., Ltd.
Shinko Building 2-12, Kanda
Ogawamachi, Chiyoda-ku
Tokyo 101, Japan

Prof. Mohamad Soerjani
Centre for Research of Human Resources
    and the Environment
University of Indonesia
Jalan Salemba 4, Jakarta 10430
Indonesia

Programme and Schedule

<u>Monday, 3rd February 1992</u>

09:30 - 10:00   Opening Session

10:30 - 12:00   <u>Session I:</u>   "Key-Note Speech"

                                   by   Prof. Arata Ichikawa
Department of Urban Engineering
Faculty of Engineering
University of Tokyo

13:30 - 15:00   <u>Session II:</u>   "Global Environmental Problems and Asian Development Patterns"

                                   by   Prof. Ryuichiro Matsubara
College of Arts and Sciences
University of Tokyo

15:30 - 17:00   <u>Session III:</u> "Hydrological Effects of Deforestation and Alternative Land Uses: A French Experiment in the Amazonian Rain Forest"

                                   by   Dr. Jean-Marie Fritsch
Senior Hydrologist
French Institute for Scientific Research and Cooperation for Development (ORSTOM)

<u>Tuesday, 4th February</u>

09:00 - 10:30   <u>Session IV:</u> "Protection of Japanese Forest and Transfer of the Japanese Experience"

                                   by   Dr. Takehiko Ohta
Forest Hydrology & Erosion Control Engineering
Department of Forestry
University of Tokyo

11:00 - 12:30   Country Paper Presentation (I)

14:00 - 15:30   <u>Session V:</u>   "Tropical Forest Potentials, Problems, and Management Efforts"

                                   by   Prof. Mohamad Soerjani
Centre for Research of Human Resources and the Environment
University of Indonesia

16:00 - 17:30   Country Paper Presentation (II)

<u>Wednesday, 5th February 1992</u>

09:00 - 10:30	<u>Session VI</u>:	"Recycling Technology of Wasted Paper: Achievements and Tasks"

by Mr. Taiji Ohashi
      Director Products Support Unit
      Fuji Xerox Office Supply Co., Ltd.

10:45 - 12:15   <u>Session VII</u>: "Regeneration of the Tropical Rainforest on New Britain Island, Papua New Guinea"

by Mr. Seiji Mori
      Project Coordinator
      Afforestation Project Promotion Office
      New Business
      Nissho Iwai Corporation

13:30 - 15:00   Country Paper Presentation (III)

15:30   Leave for Nikko

<u>Thursday, 6th February</u>

Observational Visit to Ashio: Afforestation Project: History Present Status, Problems and Future Prospects

<u>Friday, 7th February</u>

09:00   Return to Tokyo

15:00 - 16:50   Summing-Up and Final Discussion

Asian Productivity Organization

4-14, AKASAKA 8-CHOME
MINATO-KU, TOKYO
107 JAPAN
TEL.: (03) 3408-7221
TELEFAX: (03) 3408-7220
CABLE: APOFFICE TOKYO
TELEX: APOFFICE J26477

(800. 4. 1993)

Environment and Forestry Management

1993
Asian Productivity Organization
Tokyo

Report of APO Study Meeting on New Trends
in Environmental Management
(STM-03-92)

ISBN: 92-833-2124-3